International Conference on AeroSpace Science and Technology (ICASST)

Journal of Physics: Conference Series Volume 2475

Mexico City, Mexico
28 November - 2 December 2022

ISBN: 978-1-7138-7225-2
ISSN: 1742-6588

Printed from e-media with permission by:

Curran Associates, Inc.
57 Morehouse Lane
Red Hook, NY 12571

CURRAN ASSOCIATES INC. proceedings.com

Some format issues inherent in the e-media version may also appear in this print version.

Institute of Physics
Dirac House, Temple Back
Bristol BS1 6BE UK

Phone: 44 1 17 929 7481
Fax: 44 1 17 920 0979

techtracking@iop.org

Additional copies of this publication are available from:

Curran Associates, Inc.
57 Morehouse Lane
Red Hook, NY 12571 USA
Phone: 845-758-0400
Fax: 845-758-2633
Email: curran@proceedings.com
Web: www.proceedings.com

TABLE OF CONTENTS

Journal of Physics: Conference Series **2475** (2023) 011001 doi:10.1088/1742-6596/2475/1/011001

International Conference on AeroSpace Science and Technology (ICASST) 2022

J J Hernández-Gómez*[,1,2], M G Orozco-del-Castillo*[,2,3], F Ramírez-López[1,4] and C Couder-Castañeda[1]

[1] Instituto Politécnico Nacional, Centro de Desarrollo Aeroespacial. Belisario Domínguez 22, Centro, Cuauhtémoc, Ciudad de México, 06610, México.
[2] Association for the Advancement of Artificial Intelligence Student Chapter Mexico (AAAIMX). Periférico de Mérida Lic. Manuel Berzunza, Zona Dorada, Mérida, Yucatán, 97219, México.
[3] Tecnológico Nacional de México/IT de Mérida, Departamento de Sistemas y Computación. Av. Tecnológico Km. 4.5 S/N, Mérida, 97118, México.
[4] Instituto Politécnico Nacional, Unidad Profesional Interdisciplinaria en Ingeniería y Tecnologías Avanzadas. Avenida Instituto Politécnico Nacional No. 2580, Col Barrio la Laguna Ticomán, Gustavo A. Madero, Ciudad de México, 07340, México.

*Corresponding author

E-mail: jjhernandzgo@ipn.mx

1. Introduction

Welcome to the publication of the Proceedings of the *International Conference on Aerospace Science and Technology (ICASST) 2022*, which was hosted at the facilities of Centro de Desarrollo Aeroespacial (CDA), Instituto Politécnico Nacional (IPN), in the historical centre of Mexico City, Mexico, from November 28th to December 02nd, 2022. ICASST 2022 was a scientific forum for the dissemination of basic research and/or technological development projects, which allows interaction between professionals, scholars and industrialists, promoting scientific-technological collaborations in the aerospace field and related areas. The program included presentations, keynote lectures given by widely recognized national and international experts in their area, as well as courses/workshops, particularly intended for undergraduate students. The Conference was held in a hybrid way, face-to-face/virtual. Beside the academic face of this Conference, another fundamental purpose for its celebration is to commemorate the 10th anniversary of Centro de Desarrollo Aeroespacial (CDA) of Instituto Politécnico Nacional (IPN), a research centre focused on carrying out research on theoretical and applied science as well as on technological development, mainly in the institutional and national scenarios at Mexico, in the aerospace sector.

For the review and acceptance of research papers, a high-level Academic Committee was consolidated, consisting of 29 institutional, national and international researchers. Research centres and schools of the Instituto Politécnico Nacional participated in the Academic Committee, such as Centro de Desarrollo Aeroespacial (CDA), Unidad Profesional Interdisciplinaria en Ingeniería y Tecnologías Avanzadas (UPIITA), Centro de Estudios Científicos y Tecnológicos 9 "Juan de Dios Bátiz" (CECyT 9), Escuela Superior de Ingeniería Mecánica y Eléctrica Unidad Zacatenco (ESIME Zacatenco) as well as Centro de Investigación en Computación (CIC). In the national face, we had the participation in the Academic Committee of Instituto Tecnológico de Mérida of Tecnológico Nacional de México (ITM/TecNM), Centro de Investigación Científicas de Yucatán (CICY), Instituto Mexicano del Petróleo (IMP), Centro de Investigación en Ciencias de Información Geoespacial, A.C. (CentroGeo) of CONACyT, Centro de Investigación en Matemáticas, A.C. (CIMAT), Universidad Autónoma de Yucatán (UAdY), Instituto de Geofísica and Facultad de Ciencias of Universidad Nacional Autónoma de México (UNAM), Facultad de Ingeniería of Universidad Autónoma de San Luis Potosí (UASLP), as well as Universidad de Guanajuato. Finally, the international presence came hand in hand with Universidad de Córdoba in Colombia, Universidad Complutense de Madrid in Spain, Universidad de Costa Rica, and Kyushu Institute of Technology in Japan. In this way, a rigorous double-blind peer-review process was carried out, which adhered to the best international scientific review practices. Of the total of papers received, there was a 31.2% rejection rate for manuscripts that did not meet the high standards of the peer-review process carried out. Accepted papers were presented orally and subsequently published in the Journal of Physics: Conference Series of the Institute of Physics (IOP) of Great Britain, in an Open Access base, for their worldwide dissemination.

Finally, we would like to give a special mention to Secretaría de Investigación y Posgrado (SIP) of Instituto Politécnico Nacional (IPN), México, for the partial financial support received to organise the *International Conference on AeroSpace Science and Technology 2022.*

2. Justification

The new normal imposed as the general outcome of the about two years of pandemics has drastically modified the ways in which we live and interact. It has had a particularly dramatic impact in academic activities such as seminars and conferences. In this way, although we have had somehow recovered from the health emergency that COVID-19 represented, the taught lesson is that the future of many scholar activities is to remain in hybrid mode, as it aids to get to a broader international audience, to be able to receive works and speakers from any place worldwide as well as to cheapen costs and taking care of the environment.

Bearing this lesson in mind, the Committee of ICASST 2022 decided to keep the Conference in hybrid mode. This allowed us to have one keynote speaker in virtual mode directly from Japan, as well as to receive peer-reviewed works from Querétaro, Mexico as well as from Colombia, which were also presented virtually.

Furthermore, the virtual face of the Conference allowed to speakers, researchers and students fo disseminate the progress and results of their investigations in the aerospace field to the international scientific community, making the *International Conference on AeroSpace Science and Technology* a forum with a large significance and impact on the field, both regionally and internationally.

Furthermore, as part of the dissemination and divulgation activities of the aerospace sciences and technological development, this forum is fundamental for Centro de Desarrollo Aeroespacial (CDA), as is our institutional duty to perform such activities within the legal frame that constitutes us. We hope that ICASST will remain in the future as a strong motivation for the national community, students, professors, researchers and professionals, to foster their research actitivies in the aerospace and related fields, as well as this publication to share internationally and in an Open Access basis, their results.

3. Results

In the 2022 edition of the *International Conference for AeroSpace Science and Technology*, which took place from November 28th to December 02nd, 2022, were received 16 scientific works from various institutions: Instituto Politécnico Nacional (IPN) (Centro de Desarrollo Aeroespacial (CDA); Escuela Superior de Ingeniería Mecánica y Eléctrica Unidad Zacatenco (ESIME Zacatenco), Unidad Profesional Interdisciplinaria en Ingeniería y Tecnologías Avanzadas

(UPIITA), Centro de Vinculación y Desarrollo Regional Cancún (CVDR Cancún), Centro de Innovación y Desarrollo Tecnológico en Cómputo (CIDETEC) and Escuela Superior de Física y Matemáticas (ESFM)), Universidad Nacional Autónoma de México (UNAM) (Instituto de Investigación en Materiales (IIM) and Facultad de Ciencias (FC)), Instituto Mexicano del Petróleo (IMP), Centro de Investigación y de Estudios Avanzados (CINVESTAV) del IPN and Universidad de Córdoba, Colombia.

Approximately 36 authors participated in the Conference, which had about 100 persons attending face-to-face and about 200 in virtual mode.

The academic program consisted of 9 keynote speakers as well as oral sessions with the works that were submitted for the peer-review process and that were accepted for presentation at the Conference. For these purposes, seven thematic sessions were established, considering the specific topics of the received works as well as the fields of expertise of the keynote speakers:

(i) Aerospace structural and propulsion engineering,

(ii) Aerospace materials science and engineering,

(iii) Aerospace telecommunications,

(iv) Systems engineer,

(v) Training of highly specialized human resources,

(vi) Hardware development in the space field, and

(vii) Remote perception.

Moreover, courses and workshops mainly focused on undergraduate students were lectured. Although they were mainly conceived to be taken in the face-to-face mode, they were successful and the face-to-face availability quota was sold out. It is important to remark that, unexpectedly, we also received high-school students who took the courses/workshops.

In the following lists, details about the program can be found.

3.1. Keynote Speeches

(i) Title: **"Latest SmallSat/CubeSat activities at Kyushu Instituite of Technology (KyuTech)"**
Keynote Speaker: Prof. FUSE Tetsuhito
Affiliation: Kyushu Institute of Technology (KyuTech). Japan

(ii) Title: **"Orientation Control System compatible with the CubeSat standard"**
Keynote Speaker: Dr. Hugo Rodríguez Cortés
Affiliation: Centro de Investigación y de Estudios Avanzados (CINVESTAV). Mexico.

(iii) Title: **"Aerospace and technological developments at IPN"**
Keynote Speaker: Dr. Mario Alberto Mendoza Bárcenas
Affiliation: Centro de Desarrollo Aeroespacial (CDA). Mexico

(iv) Title: **"Space Development Opportunities in Mexico"**
Keynote Speaker: Dr. Carlos Roberto de Jesús Duarte Muñoz
Affiliation: Agencia Espacial Mexicana (AEM). Mexico.

(v) Title: **"Research in CECyT 9 high school: seedbed of researchers in the aerospace field"**

Keynote Speaker: M. Sc. Irving Eleazar Pérez Montes
Affiliation: Centro de Estudios Científicos y Tecnológicos No. 9 "Juan de Dios Bátiz" (CECyT 9). Mexico.

(vi) Title: **"IPNSat1 satellite project"**
Keynote Speakers: Dr. David Muñoz Rodríguez and Dr. Benjamín Varela Orihuela
Affiliation: Instituto Tecnológico de Estudios Superiores de Monterrey (ITESM) and Centro de Desarrollo Aeroespacial (CDA). Mexico

(vii) Title: **"Remote perception of Nanosatellites: Challenges and Opportunities"**
Keynote Speaker: Dr. Julio César Rolón Garrido
Affiliation: Centro de Investigación y Desarrollo de Tecnología Digital (CITEDI). Mexico.

(viii) Title: **"Remote sensing of the Earth surface: instruments and methods"**
Keynote Speaker: Dr. Juan Carlos Validivezo Navarro
Affiliation: Centro de Investigación en Ciencias de Información Geoespacial, A.C. (CentroGeo). Mexico

(ix) Title: **"Design of low-thrust trajectories with Evolutionary Algorithms"**
Keynote Speaker: MSc. Mario Augusto Ramírez Morales
Affiliation: Centro de Innovación y Desarrollo Tecnológico en Cómputo (CIDETEC). Mexico.

3.2. Courses and workshops

(i) Title: **"Newtonian physics behind orbital mechanics"**
Imparts: Dr. Jorge Javier Hernández Gómez
Modality: Face-to-face
Focus: Theoretical
Level: For undergraduate students

(ii) Title: **"Reliability of an Aerospace System"**
Imparts: Dr. Diego A. Padilla P.
Modality: Face-to-face
Focus: Theoretical and practical
Level: For undergraduate students

(iii) Title: **"Environmental tests normativity for small satellites - CubeSat"**
Imparts: Eng. Arturo Benjamín Hurtado Pérez
Modality: Face-to-face
Focus: Theoretical
Level: For undergraduate students

(iv) Title: **"Introduction to STK (Systems Tool Kit)"**
Imparts: M. en C. Emilio Noé Jiménez Navarro
Modality: Face-to-face
Focus: Practical
Level: For undergraduate students

There were seven oral thematic sessions, in which the speakers presented 11 accepted works. The time allotted to speakers to exhibit was:

- Keynote speakers were given 1:30 hours, including Q&A.

Journal of Physics: Conference Series **2475** (2023) 011001 doi:10.1088/1742-6596/2475/1/011001

- Speakers in oral sessions were allotted 30 minutes, including Q&A.

Beside the coffee breaks given to speakers and attendants, each day a social meeting was held in order to foster the interaction between authors, keynote speakers, professors, researchers as well as students.

Finally, it is important to remark that in the face-to-face event, we received a large portion of high-school, undergraduate and graduate students, which allowed them to acquire deep insights into research as well as into the aerospace field, fostering scientific vocations in the youth as well as future researchers.

4. Hybrid face of ICASST22

The keynote speeches as well as the oral thematic sessions were transmitted live at Zoom platform, and they were recorded in Full HD for their post-processing and are available upon request.

Additionally, we managed three social networks profiles for ICASST 2022, Facebook, Twitter and LinkedIn, in which the activities of the Conference were deeply diffused. They were also used to perform some divulgation of the state of the art in the aerospace sciences and technology development.

Finally, but not least important, is the ICASST 2022 website,

`https://www.icasst.org/` ,

where the full information of the event was presented, and which has been intensely updated before, during as well as after the event.

The general program, information about keynote speakers, the instructions for registration, payment, sending receipts, the guide formats for written works, updated Call for Papers (CFP), payment and registration to the Conference, a photo gallery as well as testimonials, among others, is presented in such a webpage.

The website will remain open until the next Conference, while the 2022 version will be moved and will remain accessible in a folder www.icasst.org/2022.

5. Conference Committee

The complete Conference Committee was divided as follows:

5.1. Local Committee

- Luis Fernando Rodríguez Domínguez
  Centro de Estudios Científicos y Tecnológicos 9 "Juan de Dios Bátiz", Instituto Politécnico Nacional. Mexico.
  Escuela Superior de Cómputo, Instituto Politécnico Nacional. Mexico.

- Arturo Ibsen Vargas Rosas
  Centro de Estudios Científicos y Tecnológicos 9 "Juan de Dios Bátiz", Instituto Politécnico Nacional. Mexico.

- Arturo Benjamín Hurtado Pérez
  Sección de Estudios de Posgrado e Investigación, Unidad Profesional Interdisciplinaria en Ingeniería y Tecnologías Avanzadas, Instituto Politécnico Nacional. Mexico.

- Alfredo Sánchez Martínez
  Escuela Superior de Cómputo, Instituto Politécnico Nacional. Mexico.

- Fabián Ramírez López
  Centro de Estudios Científicos y Tecnológicos 9 "Juan de Dios Bátiz", Instituto Politécnico Nacional. Mexico.
  Unidad Profesional Interdisciplinaria en Ingeniería y Tecnologías Avanzadas, Instituto Politécnico Nacional. Mexico.

- Abraham de Jesús Pablo Sotelo
  Escuela Superior de Ingeniería Mecánica y Eléctrica Azcapotzalco, Instituto Politécnico Nacional. Mexico.

- Karen Jocelyn Medina Gutiérrez
  Centro de Estudios Científicos y Tecnológicos 9 "Juan de Dios Bátiz", Instituto Politécnico Nacional. Mexico.

- Alma Diana Herrera Ortiz
  Sección de Estudios de Posgrado e Investigación, Unidad Profesional Interdisciplinaria en Ingeniería y Tecnologías Avanzadas, Instituto Politécnico Nacional. Mexico.

- Daniel Lemuel Sánchez Cabadas
  Unidad Profesional Interdisciplinaria en Ingeniería y Tecnologías Avanzadas, Instituto Politécnico Nacional. Mexico.

- Mauricio Gutiérrez Montor
  Unidad Profesional Interdisciplinaria en Ingeniería y Tecnologías Avanzadas, Instituto Politécnico Nacional. Mexico.

- Gael Emiliano Casillas Aviña
  Centro de Estudios Científicos y Tecnológicos 9 "Juan de Dios Bátiz", Instituto Politécnico Nacional. Mexico.

- Josué Enrique García Ayala
  Escuela Superior de Física y Matemáticas, Instituto Politécnico Nacional. Mexico.

- Sergio José Ramírez Gomar
  Escuela Superior de Ingeniería Mecánica y Eléctrica Zacatenco, Instituto Politécnico Nacional. Mexico.

5.2. International Scientific Committee

- Dr. Miguel Félix Mata Rivera
  Unidad Profesional Interdisciplinaria en Ingeniería y Tecnologías Avanzadas, Instituto Politécnico Nacional. Mexico.

- Dr. Mauricio Gabriel Orozco del Castillo
  Tecnológico Nacional de México/Instituto Tecnológico de Mérida. Mexico.

- Dr. José Carlos Ortiz Alemán
  Unidad de Ciencia del Agua, Centro de Investigación Científica de Yucatán. Mexico.

- Dr. Jorge Javier Hernández Gómez
  Centro de Desarrollo Aeroespacial, Instituto Politécnico Nacional. Mexico.

- Dr. Carlos Couder Castañeda

Centro de Desarrollo Aeroespacial, Instituto Politécnico Nacional. Mexico.

- Dr. Isaac Medina Sánchez
  Centro de Desarrollo Aeroespacial, Instituto Politécnico Nacional. Mexico.
- Dr. Diego Alfredo Padilla Pérez
  Centro de Desarrollo Aeroespacial, Instituto Politécnico Nacional. Mexico.
- Dr. Alfredo Trujillo Alcántara
  Instituto Mexicano del Petróleo. Mexico.
- Dr. Juan Carlos Valdiviezo Navarro
  Centro de Investigación en Ciencias de Información Geoespacial, Consejo Nacional de Ciencias y Tecnología. Mexico.
- Dr. Rodolfo David Fallas Soto
  Universidad de Costa Rica. Costa Rica.
- CSE. Sergio Ramírez Espinosa
  Centro de Estudios Científicos y Tecnológicos 9 "Juan de Dios Bátiz", Instituto Politécnico Nacional. Mexico.
- MSc. Erika Virginia de Lucio Rodríguez
  Escuela Superior de Ingeniería Mecánica y Eléctrica Zacatenco, Instituto Politécnico Nacional. Mexico.
- Dr. Mario Eduardo Rivero Ángeles
  Centro de Investigación en Cómputacion, Instituto Politécnico Nacional. Mexico.
- Dr. Alberto Luviano Juárez
  Unidad Profesional Interdisciplinaria en Ingeniería y Tecnologías Avanzadas, Instituto Politécnico Nacional. Mexico.
- Dra. Laura Ivoone Garay Jiménez
  Unidad Profesional Interdisciplinaria en Ingeniería y Tecnologías Avanzadas, Instituto Politécnico Nacional. Mexico.
- MSc. Miguel Sánchez Meraz
  Escuela Superior de Ingeniería Mecánica y Eléctrica Zacatenco, Instituto Politécnico Nacional. Mexico.
- Dr. Juan Antonio Recio García
  Universidad Complutense de Madrid. Spain.
- Dr. Francisco Javier Hernández López
  Centro de Investigación en Matemáticas A.C. Mexico.
- Dr. Alejandro Arturo Castillo Atoche
  Universidad Autónoma de Yucatán. Mexico.
- Dr. Jaime Humberto Urrutia Fucugauchi
  Instituto de Geofísica, Universidad Nacional Autónoma de México. Mexico.

5.3. Editorial Committee

Instituto Politécnico Nacional, Centro de Estudios Científicos y Tecnológicos No. 9 "Juan de Dios Bátiz". Mexico.
Instituto Politécnico Nacional, Unidad Profesional Interdisciplinaria en Ingeniería y Tecnologías Avanzadas. Mexico.

5.4. Chairs

- Dr. Jorge Javier Hernández Gómez
  General Chair
  Instituto Politécnico Nacional, Centro de Desarrollo Aeroespacial. Mexico.

- Dr. Mauricio Gabriel Orozco del Castillo
  General Co-chair
  Tecnológico Nacional de México/Instituto Tecnológico de Mérida. Mexico.

- CSE. Sergio Ramírez Espinosa
  Scientific Education Chair
  Instituto Politécnico Nacional, Centro de Estudios Científicos y Tecnológicos 9 "Juan de Dios Bátiz". Mexico.

- Eng. Victor Manuel Lechuga Gerónimo
  Student Outreach Chair
  Centro de Investigación y de Estudios Avanzados, Instituto Politécnico Nacional. Mexico.

- Luis Fernando Rodríguez Domínguez
  Student Chair
  Instituto Politécnico Nacional, Escuela Superior de Cómputo. Mexico.
  Instituto Politécnico Nacional, Centro de Estudios Científicos y Tecnológicos 9 "Juan de Dios Bátiz". Mexico.

Partners/sponsors

In addition to the Institute of Physics (IOP), ICASST22 counted as technological allies in the private sector with:

- The International Telecommunication Union (ITU),
- Jarkol Technologies S.A. de C.V.,
- The Association for the Advancement of Artificial Intelligence International Student Chapter Mexico (AAAIMX),
- The Interdisciplinary Association in Telematics and Computing A.C. (ANTACOM),
- PIGRA commercial group S.A. de C.V,
- Hidroinformática S.A. de C.V., and
- PIMAS S.A.S.

Acknowledgments

Chairs of *International Conference on AeroSpace Science and Technology* acknowledge Centro de Desarrollo Aeroespacial (CDA) of Instituto Politécnico Nacional (IPN) for its support on organizing the event, Eng. Christian Gabriel Espinosa Chávez and Centro Memorial "Eugenio

Méndez Docurro" of IPN for the use of the facilities as well as the furnishings, as well as Dirección de Investigación (DI), Secretaría de Investigación y Posgrado (SIP), Instituto Politécnico Nacional (IPN), for the partial funding of ICASST 2022 through grant 001 of the "Asistencia u organización de eventos académicos, científicos, tecnológicos o de innovación" 2022 announcement.

Peer Review Statement

All papers published in this volume have been reviewed through processes administered by the Editors. Reviews were conducted by expert referees to the professional and scientific standards expected of a proceedings journal published by IOP Publishing.

- **Type of peer review:** Double Anonymous
- **Conference submission management system:** EasyChair
- **Number of submissions received:** 16
- **Number of submissions sent for review:** 16
- **Number of submissions accepted:** 11
- **Acceptance Rate (Submissions Accepted / Submissions Received × 100):** 68.8
- **Average number of reviews per paper:** 3
- **Total number of reviewers involved:** 26
- **Contact person for queries:**
  Name: Dr. Jorge Javier Hernández-Gómez
  Email: contact@icasst.org
  Affiliation: Instituto Politécnico Nacional / Centro de Desarrollo Aeroespacial

Journal of Physics: Conference Series **2475** (2023) 012001 doi:10.1088/1742-6596/2475/1/012001

Design and testing of an electromechanical engine for a small solid-propellant rocket using non-metal materials

C V Olivar[1] and J L del Rio[2]

[1]Centro de Investigación y de Estudios Avanzados del Instituto Politécnico Nacional, Departamento de Ingeniería Eléctrica, Sección Mecatrónica. Av Instituto Politécnico Nacional 2508, San Pedro Zacatenco, Gustavo A. Madero, C.P. 07360 CDMX, México
[2]Facultad de Ciencias, UNAM. Av. Universidad 3000, Circuito Exterior s/n Alcaldía Coyoacán, CP 04510 Ciudad Universitaria, CDMX, México

E-mail: claudia.olivar@cinvestav.mx, victoriaolivardx@gmail.com (C V Olivar), jluisdrv@ciencias.unam.mx (J L del Río)

Abstract. In this paper an engine prototype of 2 degrees of freedom from a work still in progress that helps a solid fuel rocket to stay vertical during motor operation is presented. One of the philosophies of this paper consists in using cheap and accessible materials, as well as avoiding the use of industrial machinery for its manufacture. This paper describes the procedure to generate solid fuel from a mixture of sugar and potassium nitrate. Subsequently, the design of the engine and tests conserning reusability with the prototype are presented. The design is based solely on obtaining the torques required to keep the prototype rocket moving in a vertical direction. The corresponding equations are written and after linearization, the transfer function is presented and the system is simulated using a PID control. The electromechanical system (ES) as well as the solid fuel motor were manufactured, and even though the ES is not coupled to the motor so the control actions are not evaluated, the performance of the rocket motor is measured using different design options and the propulsion times were evaluated. To measure the orientation and position the BMP280 and MPU6050 sensors were selected, and their stabilization time after turning them on was evaluated. It was found that the materials can be reused at least once for tests up to 23 grams of solid fuel, and that the selected electronic devices must have a stabilization time before ignition.

1. Introduction

Nowadays, machines that manage to stabilize themselves by their own means are of great interest due to their multiple applications, one of the best examples is for aeronautics and in particular, for space exploration. This is why there has been a growing interest in the development of mobile rocket nozzles or motors[1]. In this work, we sought to obtain a prototype that would help the projectile to remain vertical. Additionally, it is common for projects that involve rocketry to have a high cost of resources[10] and they need specialized machinery. This can be seen, for example, in the theses by Riveros and Rodriguez [2], where they had to use a lathe to make the gas pressure chamber. That said, the purpose of this work was to find alternative materials and procedures that did not require very expensive and specialized machinery to develop an electromechanical motor for a small solid-fuel rocket. For this purpose, Arduino technology, low-cost pressure and movement sensors, 3D printing in PLA, and materials such as PVC and sodium bentonite were used.

The use of PLA instead of other materials was considered because it is the easiest material to use for 3D printing, since it does not require a gas recovery chamber, it is non-toxic and it is biodegradable. In addition, it was theorized that there would not be much difference between using PLA or ABS because although the latter supports higher temperatures [18], the difference is not comparable to the magnitude of the temperature of the gases ejected by the rocket.

2. Metodology

The methodology of this work is established in three parts: the first refers to the elaboration of the fuel, the second to the design of the engine and the tests that were carried out with the prototype and the propellant, and finally the tests with the electronic devices.

2.1. Solid fuel processing

There are different types of propellant for rocket and although the most efficient are usually liquid fuels, using a liquid fuel represents a greater challenge since its implementation requires more mechanical components, for this reason, solid fuel was chosen for this work, and it was made of a mixture of 35% sugar and 65% potassium nitrate [9] since this composition is frequently used for rocketry projects and has been studied and characterized in recent years[5].

It is relevant to comment that before starting, it is fundamental to follow the security measures described in appendix A of this work. To prepare the solid propellant, the following steps were performed:

i. With a scale, the masses of KNO_3 and $C_{12}H_{22}O_{11}$ to be used are weighed with the aforementioned proportion.
ii. Both elements are placed in an airtight plastic container and mixed by rotating the container at a speed of 1 rev/s for at least 2 hours.
iii. A non-stick frying pan is placed on an electric grill and heated until it reaches approximately 170ºC, which is the melting temperature of the mixture [4].
iv. Once the pan has reached the desired temperature, the mixture is poured and additionally water is added in a proportion of 10% of the total mass of the mixture, with the purpose of integrating the ingredients more efficiently.
v. With a wooden or plastic spatula, stir the mixture while continuing to heat until it begins to take on a caramel color.
vi. Immediately, the mixture is poured into cylinder-shaped containers, with a diameter slightly smaller than the internal diameter of the engine and with a height of up to 5 cm. It is important to pour the mixture into its final container soon, as this will avoid the problem of hardening of the mixture.

Tests were carried out comparing the burning time between propellants that had a hole in the center with respect to others that did not. Likewise, that they had a lid and that they did not, and experiments were also carried out varying the amount of fuel. Finally, a stainless steel metal piece was placed at the end of the motor that changes from 1/2" to 1/4" diameter, available at plumbing supply stores, to function as a nozzle. To create the hole in the center of the propellant, a glass rod is placed in the center of the cylinder when the fuel is poured. For this work, cylinders made of kraft paper with 5 layers of thickness were used.

Important notes on safety, and where to find some materials, can be found in Appendix B of this work.

2.2. Engine design and system model

To design the engine, the analysis of the problem was taken into consideration according to the scheme of figure 1 where the force of gravity, the friction force due to the air and the thrust of the engine gases due to combustion are taken into account. Disturbances due to a horizontal force of the air were neglected and the model was simplified considering that the center of pressure is above the center of mass. So that, if we consider the torques with respect to the center of mass, equation 1 is obtained:

$$\tau_{gravity} + \tau_{drag} + \tau_{motor} = \tau_{total} \quad (1)$$

Being the torques as in equations 2, 3, 4 and 5:

$$\tau_{gravity} = -gsin\theta m_2 \frac{L_2}{2} + gsin\theta m_1 \frac{L_1}{2} \quad (2)$$

$$\tau_{drag} = -\frac{1}{4} C\rho A_2 v^2 sin\theta L_2 + \frac{1}{4} C\rho A_2 v^2 sin\theta L_1 \quad (3)$$

$$\tau_{motor} = f_{motor} L_1 sinu \quad (4)$$

$$\tau_{total} = I\frac{d^2\theta}{dt^2} = I\ddot{\theta} \quad (5)$$

Where the subscripts 1 and 2 refer respectively to the lower and upper parts of the rocket taking as reference the center of mass like in the figure 1. The m refer to the masses, the L to the lengths, C to the coefficient of air friction, g to gravity, ρ the density of the air, r the radius of the rocket, A the areas of the contact surfaces, v the speed of the rocket, I the inertial matrix and the f_{motor} buoyant force of the gases[14].

Substituting equations 2, 3, 4 and 5 in equation 1, the variable $\ddot{\theta}$ can be solved as follows

$$\ddot{\theta} = \frac{\theta}{I}\{\frac{g}{2}[m_1 L_1 - m_2 L_2] + \frac{C\rho v^2}{4}[A_1 L_1 + A_1 L_1]\} + \frac{f_{motor} L_1}{I} u \quad (6)$$

thus the system has the following the transfer function

$$\frac{Y(s)}{U(s)} = \frac{f_{motor} L_1 / I}{s^2 - \{\frac{g}{2}[m_1 L_1 - m_2 L_2] + \frac{C\rho v^2}{4}[A_1 L_1 - A_2 L_2]\}/I} \quad (7)$$

Where Y(s) is the output and U(s) the angle input u. If we also do

$$\theta = x_1$$

$$\dot{\theta} = x_2$$

so that

$$\dot{x_1} = \dot{\theta} = x_2$$

$$\dot{x_2} = \ddot{\theta}$$

and if we linearize the system for small angles, the system can be expressed in matrix form 7:

$$\dot{x} = Ax + Bu \tag{8}$$

where if $\gamma = \frac{1}{I} = \frac{3}{m_1 L_1^2 + m_2 L_2^2}$ then

$$A = \begin{bmatrix} 0 & 1 \\ \gamma \left\{ \frac{g}{2} \left[m_1 L_1 - m_2 L_2 \right] + \frac{C \rho v^2}{4} \left[A_1 L_1 - A_2 L_2 \right] \right\} & 0 \end{bmatrix}$$

$$B = \begin{bmatrix} 0 \\ \gamma f_{motor} L_1 \end{bmatrix}$$

It can be verified that it is a controllable system[7].

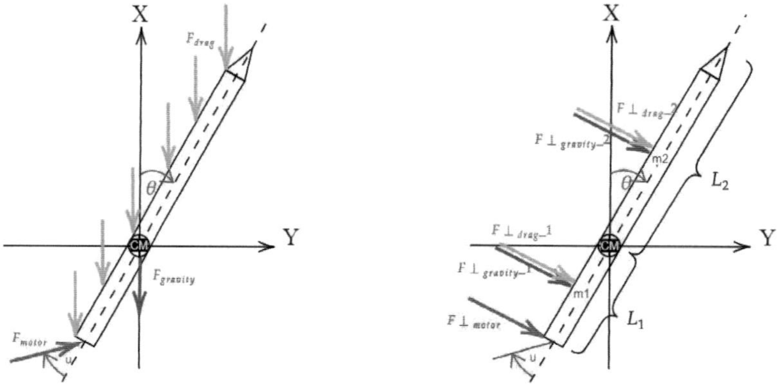

a) b)

Figure 1. Force diagrams proposed for the dynamics of the rocket. a) The force of gravity, the force of friction and the thrust force due to the gases that come out of the engine. b) The torques around the center of mass.

The purpose of modeling the system in this way is that the angle u that can be seen in figure 1 can be mechanically controlled and thus correct the orientation of the rocket so that it is again parallel to the vertical axis X. This same analysis can be considered for the perpendicular plane ZX , and after ruling out other designs for motor angle u motion control, the prototype in figure 2 was finally chosen, which is capable of moving in two independent directions.

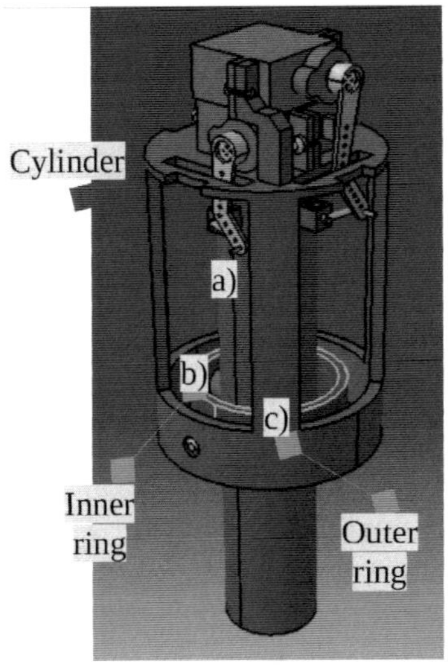

Figure 2. 3D model of the electromechanical motor prototype. The 3 main elements of the prototype can be seen: the cylinder where the propellant is located (a), the inner ring (b) and the outer ring (c).

The prototype is made up of three main pieces: the cylinder where the fuel is located, the inner ring that supports and gives the cylinder one degree of freedom, and the outer ring that supports the previous two and provides the second degree of freedom. Additionally, figure 2 shows the motors that will move the cylinder with fuel.

To carry out the resistance tests of the material, the fuel cartridges (about 12 grams of fuel) were placed inside a 1/2" PVC tube that had a lid of the same material, simultaneously half a centimeter of sodium bentonite was placed inside the PLA cylinder to serve as insulation. The PVC tube was placed inside the cylinder and in the space between the tube and the PLA cylinder bentonite was also added. Finally, the combustion was recorded on video and the time it lasted was recorded.

2.3. Evaluation of the stabilization time of both sensors

For the computer that the rocket would carry on board, three main elements were considered: the Arduino NANO microcontroller, the BMP280 sensor and the IMU MPU6050. The BMP280 sensor is capable of measuring pressure in a range from 300 to 1100 hPa, so it is possible to know the height of the rocket, while the MPU6050 has an accelerometer and an oscilloscope, so using it we can know the inclination of the rocket.

The connection of the sensors to the Arduino is shown in Table 1. In this table, two additional microcontrollers were considered, the Arduino UNO and the Arduino Mini Pro, which can have the same connections as the Arduino NANO.

Table 1. Connection of the sensors with the selected microcontroller.

BMP280		UNO/Nano/Mini Pro
	SDA	A4
	SCL	A5
	VCC	5V
	GND	GND
MPU6050		
	VCC	5V
	GND	GND
	SCL	A5
	SDA	A4
	INT	Pin 2

Table 1. Both sensors with their respective connections to the Arduino microcontroller are presented.

Once the sensors were connected, using the I2C communication protocol, reaction and stabilization tests were carried out, considering the combustion time of the propellants.

Additionally, a simulation was performed in MATLAB Simulink using two ways to control the angle θ through the angle u of the motor following the scheme of figure 3; the first one with an on-off switch that would increase u proportionally to the tilt θ of the system, and the second with a PID control whose constants were found by tuning [7].

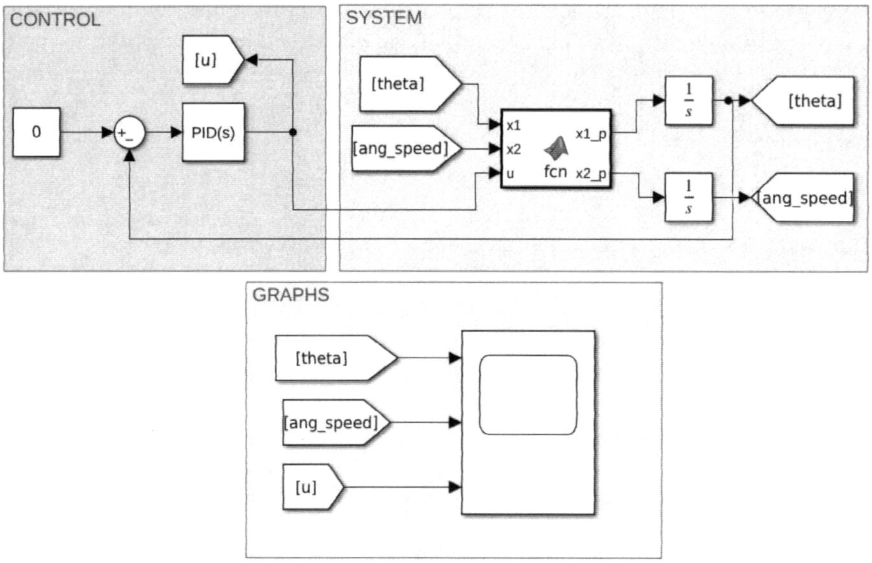

Figure 3. Simulink model of the system. The schematic for PID control.

As we can see in Figure 3, the set point is equal to 0, since that is the value we want to be. The sample time was fixed to 0.001s.

Journal of Physics: Conference Series **2475** (2023) 012001 doi:10.1088/1742-6596/2475/1/012001

3. Results and analysis

3.1. Experiments

It was found that capping propellants did not affect burn time as significantly as placing a hole in the center of propellants. This can be explained by thinking that a larger combustion surface implies more fuel to burn [4]. The tables (2- 5) express these results, were the burning time differences are noticeable.

Table 2. Example of one group of projectiles tested. The first five of them do not have lift and the second half does. Number 2 and 7 have a hole in the center.

Projectile number	Weight without fuel ($\pm 0.5gr$)	Weight with fuel ($\pm 0.5gr$)	Net fuel weight ($\pm 0.7gr^1$)
Without lift			
1	4	10	6
2*	4	10	6
3	3	12	9
4	4	12	8
5	4	12	8
With lift			
6	7	14	7
7*	6	13	7
8	7	15	8
9	7	14	7
10	7	14	7

Table 2 present 10 different tests divided in 2 groups of 5 samples and their total amount of fuel. The first half has not lift and the second half does. It is important to notice that projectiles 2 and 7 have a hole in the middle.

Table 3. Example of one group of projectiles tested. The first five of them do not have lift and the second half does. The ignition time was registered.

Projectile number	Starting time ($\pm 0.5s$)	Final time ($\pm 0.5s$)	Net ignition time ($\pm 0.7s$)
Without lift			
1	7	27	20
2*	8	15	7
3	8	27	19
4	6	29	23
5	7	29	22
With lift			
6	7	28	21
7*	4	11	7
8	4	23	19
9	15	36	21
10	17	38	21

In table 3 the ignition time of the previous samples is presented. It is very explicit how both, 2* and 7* projectiles had a shorter ignition time with respect to the others, meanwhile there is no noticeable differences between the times of ignition of both groups of projectiles.

Table 4. Example of one group of projectiles tested. The first three of them have more fuel than the second half. 14A and 12A have a hole in the center.

Projectile	Weight without fuel $\pm 0.5gr$	Weight with fuel $\pm 0.5s$	Net fuel weight $\pm 0.7gr$
14 cm of length			
14A	17	39	22
14B	14	37	23
14C	17	38	21
12 cm of length			
12A	15	33	18
12B	13	32	19
12C	15	33	18

Now, in table 4 it is presented an example of one group of projectiles with a larger amount of fuel with respect to those in tables 2 an 3. Projectiles 14A and 12A have a hole in the middle. The first three projectiles have more fuel than the second half of the group, as it can be seen in the 4th column of table 4.

Table 5. Example of one group of projectiles tested. The first three of them have more fuel than the second half. It is shown the net ignition time.

Projectile	Starting time (+/ − 0.5s)	Weight with fuel (+/ − 0.5s)	Net ignition time +/ − 0.7s
14 cm of length			
14A	7	19	12
14B	9	98	89
14C	7	55	48
12 cm of length			
12A	15	16	9
12B	13	50	43
12C	15	61	59

In table 5 the times of ignition are presented, and again, projectiles 12A and 14A had a shorter time ff ignition thanks to the hole in the center.

Likewise, it was found that adding a nozzle to the engine drastically modifies the behavior of the flow of the gases emitted by combustion, increasing the thrust they cause, which is why it is a fundamental element for what would be the final design of the rocket. Figure 4 shows a comparative image of the behavior of the gases when the nozzle is present and when it is not.

Figure 4. Comparison between the behavior of the gases when the engine does not have a nozzle (left) compared to when it does (right).

However, the weight of this nozzle was large when compared to the total weight of the engine, so to reduce its weight it was proposed to use air dry clay, with which it was possible to model a nozzle like the stainless steel one, but not tested yet.

After printing the 3D model in PLA plastic, it was confirmed that the main cylinder had two degrees of freedom as expected, however, the assembly presented some difficulties that had to be resolved with epoxy resin, likewise, it was found that the element that transmits the mechanical energy from the engine to the cylinder must be redesigned, since it is not as efficient as expected. Figure 5 shows all these elements, including those that had to be corrected with the resin.

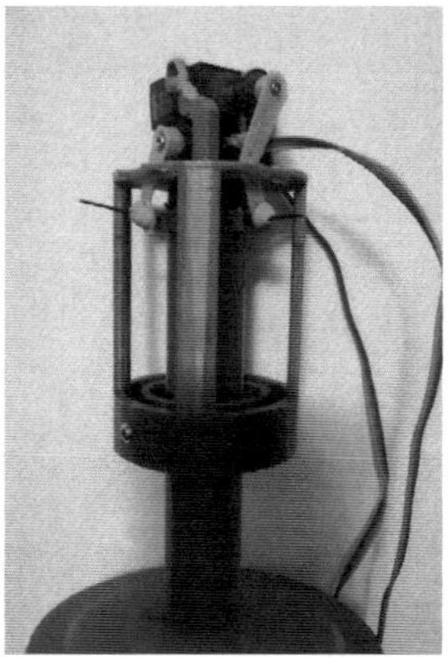

Figure 5. Printing the PLA of the prototype. For its integration, 2mm screws were used. The print lasted about 20 hours in total.

As mentioned in section 2.2, the resistance of the cylinder that contained the PVC tube with the fuel was tested to see if it was able to resist the heat and pressure exerted by the gases. It was found that in all tests it could hold up to 23 grams of fuel without a problem. Figure 6 shows the state of the cylinder after the first ignition and after its third use.

Figure 6. PLA cylinder after first ignition (left) and after use for the third time (right).

Now, regarding the MPU6050 and BMP280 sensors, it was found that for the first of them a short stabilization time should be given after initializing it before using its measurements. A minute is more than enough time for this. In the case of the BMP280 it was found that it can be used practically after turning on the computer.
The following equation was used to find the standard deviation:

$$DESVEST = \sqrt{\frac{\sum(x - \overline{x})^2}{n-1}}$$

For the MPU6050 sensor it was obtained the results from table 6 and 7:

Journal of Physics: Conference Series **2475** (2023) 012001 doi:10.1088/1742-6596/2475/1/012001

Table 6. First sampling for the MPU6050 sensor, without having discriminated the first 5 seconds of measurements..

Sample number	DESVEST Yaw	DESVEST Pitch	DESVEST Roll
1	10.42	4.65	6.87
2	10.02	3.73	5.85
3	10.37	4.63	6.87
4	9.20	2.34	4.16
5	9.19	2.34	4.16

In table 6 there are 5 samples of 5 seconds of measurements of the Yaw, Pitch an Roll angles from the MPU6050 sensor just before turning on the micro-controller in an steady position. As we can see, the standard deviation is considerable big, thus we can see that the measurements are wrong and would cause a problem when trying to compensate the inclination of the rocket prototype.

Table 7. First sampling for the MPU6050 sensor, after discriminating the first 5 seconds of measurements.

Sample number	DESVEST Yaw	DESVEST Pitch	DESVEST Roll
1	0.13	0.10	0.14
2	0.10	0.08	0.12
3	0.12	0.09	0.13
4	0.06	0.05	0.07
5	0.06	0.05	0.07

Now, in table 7 we can see immediately the huge difference with table 6 since now all of the standard deviation are less than 1, thus the angle measurements are consistent, thus for the MPU6050 it is better to wait some time for the sensor to stabilize.

For the BMP280 sensor it was obtained:

Table 8. First sampling for the BMP280 sensor, without having discriminated the first 5 seconds of measurements.

Sample number	DESVEST Temperature	DESVEST Pressure	DESVEST altitude
1	0.005	0.013	0.136
2	0.012	0.018	0.195
3	0.005	0.016	0.174
4	0.009	0.016	0.174
5	0.009	0.015	0.164

For table 8, all of its deviations are all less than 1, so the measurements are consistent even though the first measurements were not discriminated.

Table 9. First sampling for the MPU6050 sensor, after discriminating the first 5 seconds of measurements.

Sample number	DESVEST Yaw	DESVEST Pitch	DESVEST Roll
6	0.049	0.022	0.241
7	0.028	0.017	0.183
8	0.012	0.019	0.209
9	0.013	0.018	0.191
10	0.007	0.016	0.171

As we can see in table 9, there is no big difference with the previous case, so it is no necessary to wait before its use. Something important to notice as well is that the Altitude measurements have a greater standard deviation than the other two variables.

3.2. Simulation

Finally, the simulation results using the equations (2-6) can be seen in Figure 7 for a time t = 5 s and an initial condition other than 0($\theta_0 = 0.8rad$), where the angle θ is shown when u is modified in two different ways. The parameters used were g=9.8m/s, m1=m2=0.25kg, f_{motor}=144N [17], L1=0.15m, L2=0.45m, C=1.2 [15],ρ =1.2kg/m^3 [16], v=864m/s (average speed), $A1 = 2\pi r\frac{1}{2}L_1, A2 = 2\pi r\frac{1}{2}L_2$, r =0.025m.

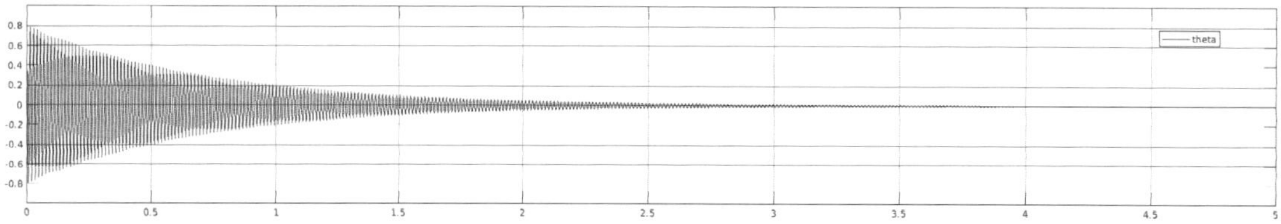

Figure 7. The simulation shows that stabilizes at 0 when u is modified according to a PID control. An oscillatory behaviour is observed , even though the system stabilizes.

Although the system modeled according to its transfer function (7) behaves as a marginally critical system[7], we can see in Figure 7 that the system manages to stabilize thanks to the control used, which is the final purpose of the engine.

4. Conclusions

As a result of this work, it was found that solid fuel works better if a hole is placed in the center to increase the combustion area, in addition to the fact that the addition of a nozzle to the engine substantially increases the thrust of the combustion gases, although due to its weight, it is necessary to look for alternatives for the material of the nozzle, such as aluminum, although it has the disadvantage that it would have to be machined on a lathe, so the option of using other materials such as air dry clay is open and thus continue the idea of a rocket model that does not require industrial machinery for its realization.

In addition even though the design of the prototype must be modified a little for the assembly of the motors, and a piece must be designed such that transmits better the mechanical movement between the motor and the cylinder, one of the main achievements of this work was to find that the selected materials are capable of being reused.

It is also important to mention that although the microcontroller used for this project was the Arduino UNO, the use of an Arduino Mini Pro should be considered since the connections with the actuators and sensors are essentially the same and it has a lower weight. It was also found that, for the electronic elements exposed in this work, it is convenient to have a stabilization time before starting the ignition and thus have adequate measurements.

Finally, according to the simulation carried out in Simulink, modifying the angle of the motor manages to stabilize the rocket if the angular velocity of the rocket is also considered, using a PID control with gains P=1, I=0.5, D=0.005.

Appendix

Appendix A: Important Safety Considerations

Two separate procedures are considered: propellant making and propellant testing. For both procedures, there must be a considerable and easily accessible body of water, a suitable fire extinguisher, preferably water-based, appropriate clothing for the occasion, a previously planned escape route and contact with the local emergency services. In the case of Mexico City, you can contact them through the telephone number 911.

During the preparation of the propellant, the heat that comes from a flame or any utensil that can generate a spark that catalyzes the chemical reaction should not be used, so metallic utensils should be avoided. It must be done in an open space, wearing protective glasses, hair tied back, and a lab coat. Fuel should not be stored for more than two days and should be kept in isolated quantities of up to 15 grams. If the mixture catches fire, you must remain calm and disconnect the electric grill from the electricity, to later pour water at the base of the fire.

Fuel tests must be done outdoors, away from flammable objects, people, and animals, with a minimum margin of 20 meters depending on the amount of fuel. Preferably, the experimental area should be delimited with a yellow tape. It is recommended to use a remote electronic device that starts combustion; it must be disconnected until the specific moment of ignition. If the device fails, it must be disconnected and wait for some time before approaching the fuel. The tests must be carried out in the company of a second observer.

Appendix B: Where to find some of the materials used in this work

Potassium nitrate with more than 90% purity can be found in specialized plant stores and markets since it can be used as a fertilizer.

Sodium bentonite shows excellent binding properties, widely used to absorb water, which is why it can be found in some brands of cat litters.

References

[1] Pimentel P 2014 *Estudio del comportamiento del cuerpo y aletas plegables de un cohete para un sistema de lanza cohetes múltiple* Master's thesis, Universidad Autónoma de Nuevo León (México)

[2] Riveros E and Rodriguez L 2010 *Diseño y construcción de un cohete aficionado controlado mediante el accionamiento de una tobera de empuje vectorial* Bachelor thesis, Universidad Militar Nueva Granada (Colombia)

[3] Málek M 2019 *Design and characterization of small solid-propellant rocket engine* Bachelor thesis, Czech Technical University (Prague)

[4] Brinley R 1960 *Rocket Manual for Amateurs* Ballantine Books pp 61-85

[5] Nakka R 2021 *Potassium Nitrate/Sucrose Propellant (KNSU)* Richard Nakka's Experimental Rocketry Web Site https://www.nakka-rocketry.net/sucrose.html

[6] A Picard, R S Davis, M Gläser and K. Fujii 2008 *Revised formula for the density of moist air (CIPM-2007) Metrology* 45 pp 149-15

[7] Katsuhiko O 2010 *Modern Control Engineering* Pearson pp 675,730-731

[8] Berggren J R 2007 *Improved Stability Analysis of Rocket Engine Nozzles* Master's thesis, Lulea University of Technology (Sweden)

[9] Gbadebo O, Inkechukwu N, Lanre M, and Olusegun S 2021 *Performance Analysis of a Dual-Fuel Sugar Based Solid Rocket Proppellant* European Journal of Engineering and Technology Research Vol. 6 No. 2

[10] Franch S 2019 *Study, implementation and test of a solid propellant rocket motor* Master's Thesis, Universitat Politècnica de Catalunya (Spain)

[11] Benson T 2022 *The Drag Equation* Glenn Research Center https://www.grc.nasa.gov/www/k-12/VirtualAero/BottleRocket/airplane/drageq.html

[12] Romero A and Herrera R 2010 *Diseño de un motor cohete de combustible sólido para un proyectil de sembrado de nubes* Bachelor Thesis, IPN Escuela Superior de Ingeniería Mecánica y Eléctrica (Mexico)

[13] Davies M 2003 *The Standard Handbook for Aeronautical and Astronautical Engineers* 1st Edition Mc Graw Hill

[14] Fox and McDonald's 1998 *Introduction to Fluid Mechanics* John Wiley and Sons 5th ed pp 444-457

[15] Bruschi G, Nishioka T, Tsang k and Wang R 2003 *A comparison of analytical Methods Drag Coefficient of a Cylinder* University of California MAE 171a https://sv.20file.org/up1/916_0.pdf

[16] Engineering ToolBox 2003 *Air - Density, Specific Weight and Thermal Expansion Coefficient vs. Temperature and Pressure* https://www.engineeringtoolbox.com/air-density-specific-weight-d_600.html

[17] Naka R 2001 *Worked example #3 Calculate the ideal thrust and thrust coefficient* http://www.nakka-rocketry.net/articles/thr_example5.pdf

[18] O'Connel J 2021 *ABS vs PLA: Comparación de filamentos para impresión 3D* https://all3dp.com/es/2/abs-vs-pla-filamento-impresion-3d-plastico/

Journal of Physics: Conference Series **2475** (2023) 012002 doi:10.1088/1742-6596/2475/1/012002

Calculation of environmental loads for a CubeSat using MATLAB

A Hernández[1], J A Ferrer[1], D Gaviria[1], C Romo[1] and R G Chávez[1]
[1]3001 Juriquilla Boulevard 76230, Querétaro, México
E-mail: alexkmex@comunidad.unam.mx, ferrerp@unam.mx, dafne.gaviria@comunidad.unam.mx, carlosrf@unam.mx, rchavez@comunidad.unam.mx

Abstract. From 2011 to 2022, around 1,663 CubeSats between 1.1 kg and 10 kg were launched worldwide. These satellites were used for sensing, technology development, communications, and scientific applications. To develop these satellites, it is necessary to determine the thermal loads for the calculation of the internal temperature distribution of the components, which allows us to design the thermal control system. In this work, the mathematical formulations are presented to be able to obtain the thermal loads. Simulations were carried out with STK which assisted in obtaining the thermal loads and another simulation was carried out with MATLAB where the results were later compared, the values to be studied being solar radiation, albedo, and planetary radiation. When comparing the results, an error of 9.32% was obtained, which is discussed in the conclusions. The importance of developing a tool that allows obtaining the thermal loads and that said tool be open source was discussed.

1. Introduction

The term CubeSat comes from the design proposed by the California Polytech[5], they proposed a new type of satellite and one of its purposes was to be a standard.

Having a standard for building something has the advantage to be easier to build and finding new pieces to work with, also having the disadvantage that it has a small volume so it can't be used for complex missions and has a small time of operation.

While CubeSat missions are aimed to be affordable, in the context of Latin American Universities they are still expensive to work with. For example, for a 1U CubeSat, a conservative budget estimate might be around $500,000[9]. This considers all expenses involved in the project. For a public university, this amount of money is almost unreachable, so they must rely on international projects like KiboCUBE[9] or government support, but the latter is harder to get.

The use of nanosatellites has increased over recent years, due to their low cost and their potential to offer comparable uses with larger satellites, due to the previously mentioned factors, countries with smaller budgets such as Mexico have been interested in this type of satellite such as the CubeSat Aztechsat-1[7] and the NanoConnect-2[14].

Table 1. Satellites launched by Mexico [9]

Name	Responsible	Year	Type	Use
-I	SEDENA	2019	3U	Military
AztechSat-1	UPAEP	2019	1U	Academia
Nano Connect-2 (SAI-1)	ICN-UNAM	2021	2U	Academia
D2/AtlaCom-1	SpaceJLTZ	2021	6U	Commercial
Painani-II	SEDENA	2021	3U	Military

Satellites of any size are made up of different components grouped into subsystems, which are designed around a payload, which is the component that is defined based on the objective of the space mission. These components operate in a temperature range called operational, and when this range is exceeded, the so-called survival range is reached, which can irreversibly affect the operation of a given component.

Due to this, the need arises to offer solutions for the dissipation and management of thermal energy through a subsystem known as thermal control, which will oversee handling sudden changes in temperature so that during the operation of the satellite operating ranges are exceeded. To do this, the following steps are followed in a summarized manner:

(1) Once the space mission has been defined, the first calculations of orbits are made to fulfill this mission.
(2) From these orbits, the external thermal loads are calculated.
(3) They are determined the payload and the elements of all the subsystems.
(4) The internal dissipations of each component are defined.
(5) The temperature distribution along the satellite is calculated through the energy equation.
(6) It is verified that the temperature ranges do not exceed the operational temperature ranges of each of the internal elements.
(7) A desired thermal management solution is obtained if the ranges are not exceeded.
(8) If the operating ranges are exceeded, other thermal control elements (passive and active) are added, and the entire process is repeated from point 5.

So, the main objective of this paper is to introduce the problem of obtaining the thermal loads a spacecraft in a low earth orbit experiences, first by using the tool STK, and by familiarizing with that tool, start to develop an own code, so in future projects, Mexican universities and students can use a free access tool that does not rely on private software.

1.1. Orbital and thermal load calculation tools
There are currently programs available to perform orbital simulation for free like GMAT, or for a fee like FreeFlyer or STK. Each of these tools has been used in real flight missions, so each one is tested and verified, so each user will have their preference, whether it is for cost, knowledge of its use, or for some other reason.

Systems Tool Kit (STK)

Systems Tool Kit (STK)[4] is a platform for analyzing and visualizing complex systems, we can use this platform to propagate the orbits of satellites and obtain data related to the orbits.

By using STK the user can:

- Select, build, or import vehicle and sensor models to represent a system
- Simulate the system proposed in real-time
- Collaborate with other platforms to create a more complete project

An additional advantage that STK offers is being able to calculate thermal loads in an accessible way. However, to access this environmental module it is necessary to have a license with a cost. This is important to mention, since, in the case of Mexico, these costs cannot be easily covered by public institutions throughout the country, so it is necessary to develop, as a country, its tools that can be used without restrictions for use. academic, non-commercial, non-governmental, non-military.

1.2. Tools for calculating temperature distribution
Once the external thermal loads have been calculated and the thermal dissipation of the internal components determined, the temperature distribution is calculated. Commercial programs such as ANSYS[3], Thermal Desktop, or ESATAN-TMS, can be fed with information on thermal loads from STK to calculate temperatures. However, there are other options with which it is possible to carry out the complete calculation of the temperature, including the thermal loads. These options are presented below:
Thermal Desktop[2]

Thermal Desktop includes all aspects of model creation. Built-in finite difference, finite element, and concentrated capacitance objects which can be combined in any configuration.
ESATAN-TMS[8]

ESATAN-TMS is the software for the complete analysis process, in its thermal analysis includes radiation analysis for aerospace vehicles, it's developed for professionals in the areas of thermal analysis so it includes all the tools necessary for the analysis.

2. Background
The use of nanosatellites in the aerospace industry has been growing continuously over the years, in the case of Mexico, this type of satellite has grown in popularity thanks to several university groups that want to create and develop knowledge for their students and thus the Mexican aerospace industry. Calculate the thermal loads, there are already programs dedicated to this as mentioned above. Although there are also other programs that we can use, the importance of having our code is not depending on an external entity and being able to adjust the code to our needs.

It was mentioned that the electronic components operate in a certain range of temperatures, that's where the thermal control subsystem operates. The following table shows the operating temperatures of some components commonly used in a CubeSat.

Table 2. Typical Operating Temperatures[18]

Component	Minimum Temperature [°C]	Maximum Temperature [°C]
Batteries	10	30
Antennas	-100	100
Solar Panels	-150	100
Reaction Wheels	-10	40

Table 2 shows typical values of the operating temperatures of certain components, these temperatures will only be for reference, since the components selected in a real mission may have different values, either with a few more degrees of tolerance or a more restrictive temperature range.

3. Heat transfer methods

The spacecraft while traveling in space is subject to sudden changes in temperature, up to a variation of 40°C in a single hour, as shown in [6] and [6], thermal gradients are caused and therefore the phenomenon of heat transfer from the environment to the satellite and vice versa is presented. In the case of a spacecraft in orbit around the earth, the most relevant modes of heat transfer are conduction and radiation. Conduction only occurs within the satellite, and radiation occurs within the satellite and outside of it. In this work, only the effects of radiation on the satellite are analyzed, so it will be called for the space environment, solar radiation, albedo, which is the solar radiation reflected by the Earth's atmosphere, and direct radiation from the Earth.

In this work, the way to calculate these effects is presented, employing two methods: the first will use the STK program, which is a program that allows analyzing the orbit of a satellite, and the second will use some equations obtained in the notes of [12]and the results of the two methods that will be compared.

It had previously been mentioned that STK is a program that is not free, although it is true that for the use that we will give it, the initial license, which is free and can be downloaded from its website, is sufficient. So, the functionality given by STK will be appropriate, when using MATLAB to perform the calculations, it is due to the way of treating data tables, which facilitates its calculation, in addition to the fact that STK and MATLAB have tools that facilitate sharing data between yes, a functionality that is expected to be used as future work.

Figure 1 shows how thermal loads affect the satellite and gives a clue as to how the thermal analysis will be performed.

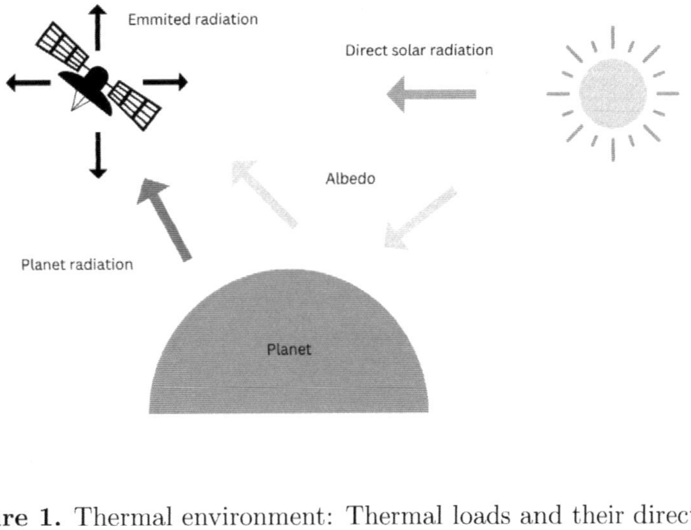

Figure 1. Thermal environment: Thermal loads and their direction

Figure 1 shows how energy is transferred to the satellite in this radiation, which is mainly modeled by the Stefan-Boltzmann Law, and for the analysis, it will be considered that our satellite is only a node without properties of its materials. Equation (1) shows the Stefan-Boltzmann Law.

$$E = \varepsilon\sigma(T_e)^4[\frac{W}{m^2}] \tag{1}$$

Being E, the power received in [W/m2], ε is the emissivity of the surface, σ the Stefan-Boltzmann constant, and Te is the temperature emitted by the surface. The way in which the external loads mentioned in figure 1 are modeled are:

$$\dot{q}_{sol} = \alpha_s G_s A_s \cos\theta \quad [W] \tag{2}$$

$$\dot{q}_{alb} = \alpha_s a G_s A_{sat} F_{sat-P} \quad [W] \tag{3}$$

$$\dot{q}_{planet} = \varepsilon_{sat} A_{sat} F_{sat-P} \varepsilon_\rho \sigma T_p^4 \quad [W] \tag{4}$$

These are the expressions in general, which will be used in the case of the equations used in the lessons of [16], in equation (2) the thermal load due to solar radiation is modeled, α_s is the absorptivity of the surface, G_s is the solar constant, which although 3 values are known, the average of 1367 [W/m^2] [13]will be used, A_{sat} is the area of the satellite that sees the solar rays directly, θ is the angle that the surface of the satellite forms with the Sun.

In equation (3), the terms α_s, G_s, A_{sat} are the same as equation (2), a is the albedo factor that for our planet has a value of 0.3,F_{sat-P} is the view factor between the Earth and the surface of the satellite, and finally in equation (4), the terms G_s, A_{sat} and F_{sat-P} are the same as equation (3), the term ε_{sat} is the absorptance of the surface of the satellite, ε_p is the emissivity of the Earth which has a value of 0.61(12), σ is the value of the Stefan-Boltzmann constant [1] the term T_p^4 is the black body temperature of the planet, which in the next chapter will be explained.

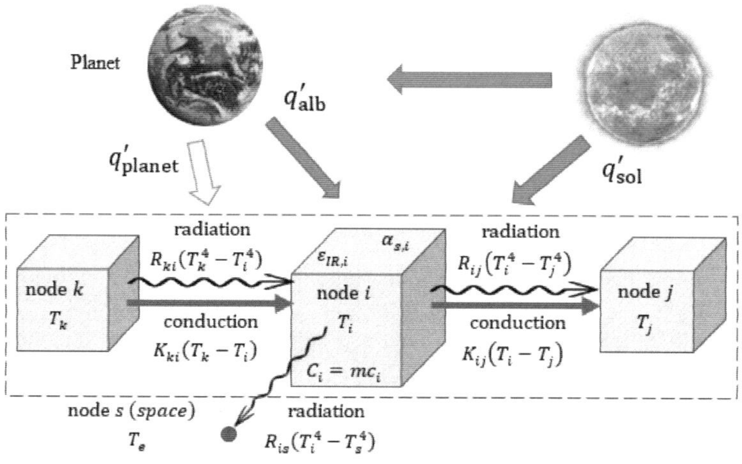

Figure 2. Thermal model of the satellite.

Figure 2 shows how the thermal model of the satellite would now also consider internal loads and heat transfer by conduction.

Journal of Physics: Conference Series **2475** (2023) 012002 doi:10.1088/1742-6596/2475/1/012002

4. Simulations

The two simulations were carried out with the MATLAB program, the first being also using the STK program to obtain orbital data, such as altitude, vectors between the satellite and the Earth and vectors between the satellite and the Sun. The second simulation it was only done using the equations in the notes of [12], these equations are calculated in MATLAB and in both cases the results are shown in graphs, the simulation time is as if the satellite gave 3 orbits to the planet. For the orbital data, the International Space Station was chosen to be our satellite, it has an altitude of 400 km, an inclination of 51.6° [1] and an orbital period of 92.68 minutes. Figure 3 shows how the simulations will be done.

The initial conditions are displayed in table 3.

Table 3. Initial conditions for both simulations

Condition	Value
Altitude	400 [km]
Solar constant	1367 [W/m2]
Orbital period	92.68 [min]
Albedo coefficient	0.3
Earth emissivity	0.6
View factor of Earth	0.8855

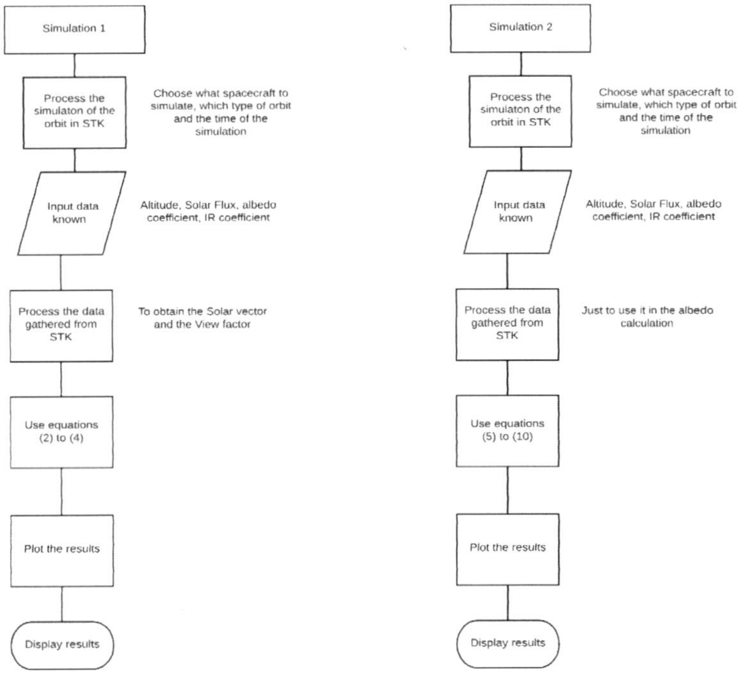

Figure 3. Procedural diagram of both simulations

4.1. First simulation
As mentioned, in the first simulation, the STK program will be used, in which the International Space Station was chosen, we choose this spacecraft because is a very known one, its conditions are very well known by numerous sources and it still has several operative years left, so if someone wants to check the values obtained in STK in the following years he will still be able. Having this vehicle, a report with the orbital values was generated, in the STK simulation as it can be seen in figure 4 values were extracted than 2 days, but only three orbits will be shown in the results.

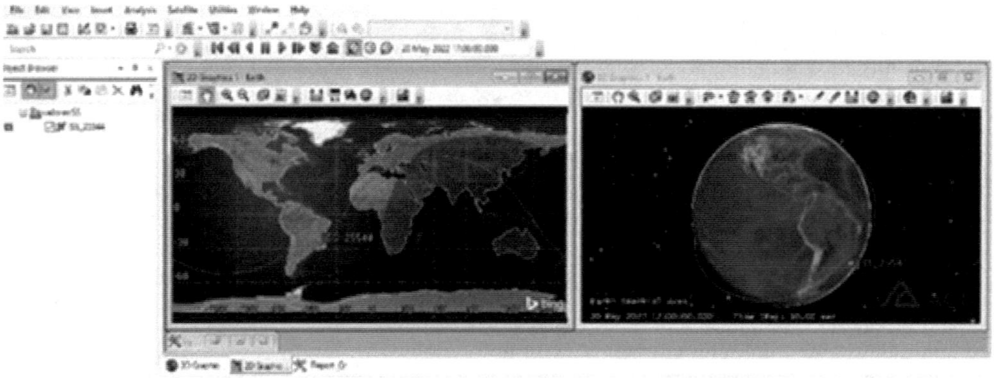

Figure 4. Simulation of the orbit of the ISS

With the report generated in STK, it was exported as an Excel file and then imported into MATLAB, with these values the data of the three external loads were obtained.

To calculate the thermal loads, we use the data obtained from STK which serve us to know when the spacecraft will be in eclipse and the view factor between the spacecraft and the Earth, by knowing those values we can use the equations (2) to (4) and have the final values of the thermal loads.

4.2. Second simulation
For the second simulation only, some values obtained in STK were used, to obtain the albedo load, these values are those of the angles between the satellite, the Earth, and the Sun, so the values of that load will be similar in the two simulations.

Solar radiation is considered to have minimal variation. It varies when the satellite is in an eclipse. To know if the satellite is in eclipse, the orbital period is first calculated and then the eclipse time. This generates a table that indicates whether the satellite is in eclipse or not.

$$T_0 = 2\pi\sqrt{\frac{(H+R)^3}{GM}} \quad [s] \tag{5}$$

With equation (5) we calculate the orbital period, where H is the altitude of the satellite, R is the radius of the Earth, which is 6370 [km], G is the universal gravitational constant with a value of $6.6710^{11}[\frac{Nm^2}{kg^2}]$ and M the mass of the Earth $5.9710^{24}[kg]$. With the equation (5) we can know how much time in seconds the spacecraft takes in doing a full orbit, this is the first step in knowing the eclipse time, so we can express the time the spacecraft is exposed to solar radiation.

Now in the next equation, (6) we calculate the eclipse time when the spacecraft is not exposed to solar radiation. This equation will give us a number, that is the fraction of time that the orbit is in eclipse, so by using equation (5) and multiplying the result by the result from equation (6) we get the time of eclipse.

$$\frac{T_e}{T_0} = \frac{1}{\pi}\arccos(\frac{\sqrt{2h+h^2}}{(1+h)\cos\beta}) \tag{6}$$

For the calculation of the terrestrial radiation, it is considered that the planet behaves isothermally, the temperature of the planet will be 288 [K] and an emissivity of 0.6, so the model is expressed by equation (7), The terms have already been explained in equation (4), just that now with the simplified model we remove the terms of the area. For equation (8) and (9) the view factor will be calculated as if the spacecraft is a small plate looking directly at the planet, the expression for getting this view factor is obtained from [15], the equation is (7):

$$F_{sat-P} = (\frac{r_e}{r_e+h})^2 \tag{7}$$

With re being the radius of the Earth, which is 6378 [km](17), and h is the altitude of the spacecraft.

$$q_{ir} = \varepsilon\sigma F_{sat-P}T_\rho^4[\frac{W}{m^2}] \tag{8}$$

For the calculation of the albedo, the simple model is first calculated, the albedo factor, ρP, being 0.3, Es the solar radiation. The simple model looks like the equation (8),

$$q_{alb0} = \rho p F_{sat-P}E_s[\frac{W}{m^2}] \tag{9}$$

Then we just have to multiply the simple model by the ratio of angles that we get from STK.

$$q_{alb} = q_{alb0}\cos\phi\cos\beta[\frac{W}{m^2}] \tag{10}$$

5. Results

5.1. Solar radiation

The three graphs of the loads obtained in the simulations are shown. The results will be displayed in pairs so the comparison between both simulations is easier to see, the graphs on the left will be of the first simulation and the graphs on the right will be of the second simulation

Figure 5 shows the heat flux due to solar radiation; it has a constant value in the parts of the orbit where it is facing the sun and then drops to zero because it stops facing the sun, in the eclipse. The difference between both results is that in the first simulation the time in eclipse is shorter, compared to the second simulation, in the proposed simulation, the solar radiation will affect longer the satellite, which could represent an increase in the temperatures inside the spacecraft.

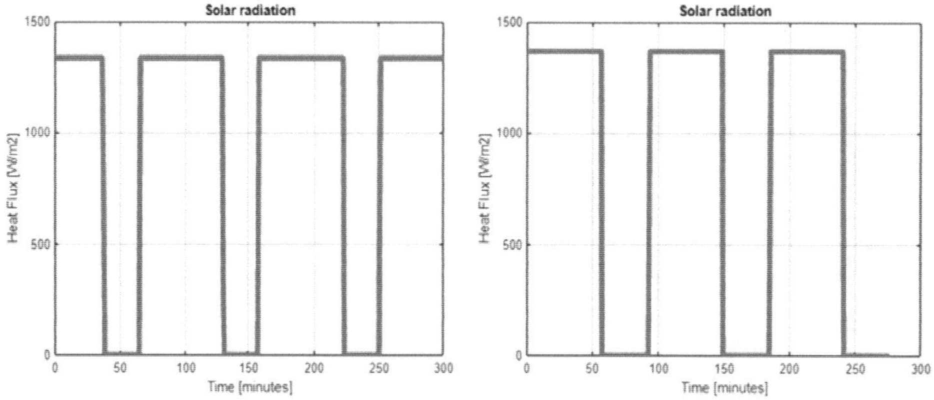

Figure 5. Graphs representing the solar radiation load

5.2. *Albedo*

Figure 6 shows the values of Albedo, this being an effect that depends on solar radiation, it varies over time because the spacecraft moves in parts where the planet is not facing the sun, so it's not reflecting the radiation.

The graphs are almost identical except that the top value, when the spacecraft is facing the planet directly and the reflected radiation is at its max, is different, in the first simulation that value is 231 $[W/m^2]$, while in the second simulation that value is of 209 $[W/m^2]$, that difference is because of how the View factor is calculated, in the first simulation the data obtained from STK offers us a view factor more similar as how the satellite would look to the planet, while in the second simulation we considered the spacecraft as a flat plaque facing directly to the planet.

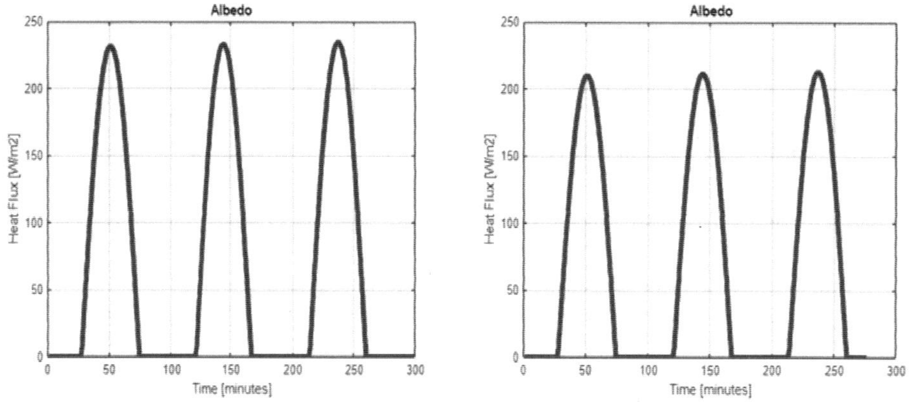

Figure 6. Graphs representing the albedo load

5.3. *Earth radiation*

Figure 7 shows the results obtained from both simulations, in the first simulation the results show a variation over time, given that STK gives values that vary over time, the values of the View factor, meanwhile in the second simulation given the consideration that the spacecraft is a small plate that always faces the planet, and that the temperature of the Earth is considered constant, that gives us a constant value over time. Considering that we assume the spacecraft is a small plate, both values are very alike.

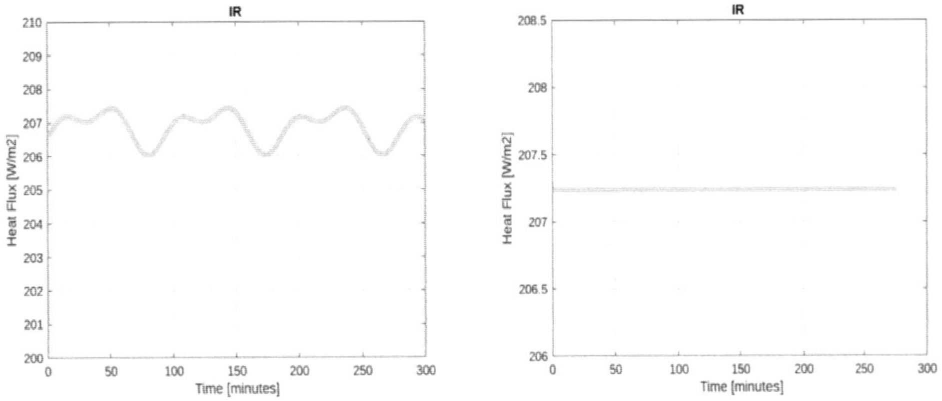

Figure 7. Graphs representing IR load

5.4. Discussion of the results

The results obtained in both simulations show that while they are approximations, they still can be used as early approximations to see what conditions the spacecraft will be exposed to. The thermal loads have heavy importance in the thermal control and design of a spacecraft, like previously mentioned in table (2), some components of the spacecraft are sensible to high or low temperatures, mainly the batteries, that have a short margin of temperature where they can properly operate, so that emphasizes the importance to properly calculate the thermal loads, these will be the alongside the energy dissipated by the electronic components, the thermal loads that apply energy to the system that is the spacecraft.

Between the results of the Earth radiation and Albedo a relative error is calculated, the equation to calculate the error is presented in equation (11).

$$e_{relative} = |\frac{theoreticalvalue - experimentalvalue}{theoreticalvalue}|x100 \tag{11}$$

We take the values of the first simulation as the theoretical value and the values of the second simulation as the experimental ones.

The relative error for the planet radiation is 0.1727% and the relative error of the albedo is 9.32%, the error in the planet radiation is small given that the main thing that varies is the view factor, and for the case of the albedo the error is also because of the view factor. So we can see the importance of having the appropriate view factor consideration, like in [6] where the difference in planet radiation is very small.

In reference [6] they perform the calculation of the external loads of a CubeSat with an altitude of 600 km, they mention that they obtained the value of the planet radiation, and their results for the hot case show a constant behavior, like the results we obtained.

In reference [17] the authors do a thermal analysis of a CubeSat, they first calculate the thermal loads, and for the Solar radiation they show several graphs of how each face of the CubeSat is getting affected. It shows the difference of a small area affects the thermal load. In [10] they design the thermal control of a CubeSat that will be at an altitude of 600 [km], they show the values of the thermal loads they will be using for their design and for the Albedo and planet radiation the values they use are very close to the ones we got. And finally, in [11] they display a graph where they show the values of the solar radiation, a graph where it shows the time of the eclipse, and also for how they consider the thermal loads affect each different face of the CubeSat, a characteristic that will be needed to considerate in future works.

6. Conclusions

In this investigation, it was possible to obtain the values of the external loads of a satellite. MATLAB was used as an analysis tool. It is important to highlight that the development of the code using MATLAB is the first step to building a free-access solution to calculate the temperature distribution of a satellite. Means that this work will not be the final result, but it's just the first step to obtaining a proper tool to use for our purposes. The calculation of the external loads will later be used to carry out the thermal analysis of a satellite and the design of the thermal control of the said satellite.

The results of this code were verified with simulations carried out from the commercial program STK, where an error of 9.32% was obtained. This variation in the results is due to the considerations made as a first approximation, mainly how the view factor is obtained. For future work, it is recommended to obtain a better approximation by varying the external loads with emphasis on the Albedo load and the eclipse times.

It is emphasized that this type of simulation supports the effort to develop CubeSat-type nanosatellites in our country, since, although as previously mentioned in the introduction, there are already several commercial programs that can perform these calculations, in a country like Mexico it is important to generate this knowledge to have a free access program for Mexicans in the future.

Acknowledgments Special thanks to the collaborators that make up the High Technology Unit, my colleagues, and our university that gives us the space to collaborate.

References

[1] . May 2022. URL: https://www.esa.int/Science_Exploration/Human_and_Robotic_Exploration/International_Space_Station/ISS_International_Space_Station.

[2] *Aertia - Thermal Desktop.* www.aertia.com. URL: http://www.aertia.com/productos.asp?pid=190 (visited on 01/15/2023).

[3] ANSYS. *Engineering Simulation 3D Design Software — ANSYS.* Ansys.com, 2017. URL: https://www.ansys.com/.

[4] *Ansys STK — Digital Mission Engineering Software.* www.ansys.com. URL: https://www.ansys.com/products/missions/ansys-stk.

[5] CalPoly. *Cubesat design specification. The CubeSat Program, California Polytechnic State .... 2009;8651.* June 2004.

[6] S. Corpino et al. "Thermal design and analysis of a nanosatellite in low earth orbit". In: *Acta Astronautica* 115 (2015), pp. 247–261. ISSN: 0094-5765. DOI: https://doi.org/10.1016/j.actaastro.2015.05.012. URL: https://www.sciencedirect.com/science/article/pii/S0094576515001940.

[7] . *El primer nanosatélite 'hecho en México' será lanzado este miércoles por la Agencia Espacial Mexicana.* El Financiero, 2019. URL: https://www.elfinanciero.com.mx/ciencia/el-primer-nanosatelite-hecho-en-mexico-sera-lanzado-este-miercoles-por-la-agencia-espacial-mexicana/ (visited on 01/15/2023).

[8] *ESATAN-TMS: Home.* www.esatan-tms.com. URL: https://www.esatan-tms.com/ (visited on 01/15/2023).

[9] Jorge Alfredo Ferrer-Pérez et al. "The Development of CubeSats in Latin America and Their Challenges on the Design of Thermal Control Systems". In: *Southern Space Studies* (2022), pp. 41–64. DOI: 10.1007/978-3-030-97959-1_3. (Visited on 01/15/2023).

[10] Soo-Jin Kang and Hyun-Ung Oh. "On-Orbit Thermal Design and Validation of 1U Standardized CubeSat of STEP Cube Lab". In: *International Journal of Aerospace Engineering* 2016 (2016), pp. 1–17. DOI: 10.1155/2016/4213189. URL: https://www.hindawi.com/journals/ijae/2016/4213189/ (visited on 10/25/2019).

[11] Róbert Kovács and Viktor Józsa. "Thermal analysis of the SMOG-1 PocketQube satellite". In: *Applied Thermal Engineering* 139 (July 2018), pp. 506–513. DOI: 10.1016/j.applthermaleng.2018.05.020. URL: https://www.sciencedirect.com/science/article/pii/S1359431117350913?casa_token=gTAMRUW0qIIAAAAA:hZ8dR8_k_AsFrgQSiUP9gp_1rxDJ2-9hU0sbyFM3U_wjnmjcD_Vj94PvBC94c0X83-LNfp4wjg (visited on 10/30/2022).

[12] Isidoro M. *Spacecraft thermal modelling and testing 1 SPACECRAFT THERMAL MODELLING AND TESTING*. webserver.dmt.upm.es, 2022. URL: http://webserver.dmt.upm.es/~isidoro/.

[13] J. Meseguer, I. Pérez-Grande, and A. Sanz-Andrés. *Thermal control design. Spacecraft Thermal Control*. June 2012.

[14] *Refrenda la UNAM su potencial científico con lanzamiento de nanosatélite al espacio*. www.dgcs.unam.mx. URL: https://www.dgcs.unam.mx/boletin/bdboletin/2021_176.html (visited on 01/15/2023).

[15] SL Rickman. *A Simplified, Closed-Form Method for Screening Spacecraft Orbital Heating Variations*. URL: https://tfaws.nasa.gov/TFAWS02/Data/Thermal/Rickman.pdf.

[16] . *SPACE VEHICLE DESIGN CRITERIA (ENVIRONMENT) C A S E F I L E C O P Y SOLAR ELECTROMAGNETIC RADIATION NATIONAL AERONAUTICS AND W A C € ADMiNISTRATlON*. URL: https://ntrs.nasa.gov/api/citations/19740019421/downloads/19740019421.pdf.

[17] E. R. Subramanian and V. Chandrasekaran. *Transient thermal analysis of a nano-satellite in low earth orbit*. manipal.pure.elsevier.com, Jan. 2012. URL: https://manipal.pure.elsevier.com/en/publications/transient-thermal-analysis-of-a-nano-satellite-in-low-earth-orbit (visited on 01/15/2023).

[18] James Richard Wertz, David F. Everett, and Jeffery John Puschell. *Space Mission Engineering: The New SMAD*. Microcosm Press, 2011. URL: https://books.google.com.mx/books/about/Space_Mission_Engineering.html?id=VmQmtwAACAAJ&redir_esc=y (visited on 01/15/2023).

Teletraffic analysis and simulation on FSO systems: the traffic generators approach

M Gutiérrez-Montor[1], G A Yañez-Casas[*,2], C E Enríquez-Ortiz[1] and J J Hernández-Gómez[2]

[1] Instituto Politécnico Nacional, Unidad Profesional Interdisciplinaria en Ingeniería y Tecnologías Avanzadas, 2580, 07340, Mexico City

[2] Instituto Politécnico Nacional, Centro de Desarrollo Aeroespacial, Belisario Domínguez 22, 06010, Mexico City

*Corresponding author

E-mail: g.yanezcasas@gmail.com

Abstract. Free Space Optical Communications (FSO) has become an interesting topic to researchers in recent years since the amount of data generated by devices is growing, and it is necessary to use data links that support high bandwidth to transmit them from one device to another. This technology is used to establish not only terrestrial links but also space links.

A CubeSat satellite is generally deployed in the Low Earth Orbit (LEO) region. To maintain the altitude, the satellite must have a high orbital speed. For that reason, the time to transmit data between a particular ground station and the CubeSat satellite is limited. On the other hand, the volume, mass, and energy storage capacity are restricted in a CubeSat. The greater the bandwidth capacity of radio frequency links, the greater the demand for volume, mass, and energy they require. For that reason, to transmit a significant amount of data, traditional radio frequency links are not suitable and are becoming replaced by FSO as technology improves.

Despite research in physical layer technologies on FSO (modulation schemes, error mitigation techniques, pointing and tracking systems), there is very little research in the literature about data link layer protocols for FSO. Secondly, there is little research to measure the data traffic demand on CubeSat satellites so that a certain data link layer protocol can be selected or adapted to be implemented in an FSO system.

This paper presents research to address the issue of teletraffic through the use of traffic generators. The result is the design and development of a traffic generator for a discrete event simulator that will later be used to observe the behavior and to measure the performance of the Selective Repeat ARQ protocol in a simulated satellite FSO link, in order to propose improvements to adapt the protocol to this scenario.

1. Introduction

To establish a link between a CubeSat satellite in Low Earth Orbit and a ground station, a lot of challenges have to be overcome. A survey about CubeSat satellites, applications, hardware subsystems and components, communication protocols, and current and future challenges are presented in [1]. This paper represents a part of a project that focuses on analyzing the behavior of data transmission protocols and flow control strategies over FSO links for CubeSat satellites.

The Consultative Committee for Space Data Systems (CCSDS) is currently working on the definition of two space FSO physical layer standards: High Photon Efficiency (HPE)

and Optical On-Off Keying (O3K) [2]. The second one defines a protocol that utilizes the Intensity Modulation/Direct Detection On-Off Keying (IM/DD OOK) modulation scheme. This modulation scheme is the most studied and used to communicate with small satellites deployed in the Low Earth Orbit, such as CubeSats.

Nevertheless, there is little research on data link protocols over FSO. In [3] it is criticized that the research to date has focused on studying error mitigation schemes in FSO links by means of physical layer techniques such as those described in [4]: forward error correction, adaptive transmission, adaptive modulation, adaptive optics, aperture averaging, spatial diversity, hybrid RF/FSO links, relay satellite assisted transmission, digital signal processing, and pointing, acquisition and tracking (PAT) systems. For this reason, [3] presents a study and recommendations on how to apply error correction schemes in FSO links through data link layer functions, that is, through retransmission protocols such as Automatic Repeat Request (ARQ) or Hybrid Automatic Repeat Request (HARQ).

In [5] and [3] the benefits of the *cross-layer design* approach to FSO links are discussed. When protocols are designed with the *cross-layer design* approach, there may be an exchange of information between the physical and data link layers, which are traditionally decoupled when using the *layered design* approach. Thanks to the exchange of information between layers, the data link protocols can have an adaptive data rate feature that responds to changes in the channel.

The article [6] discusses some ways to optimize data transmission in satellite optical links in which *Stop-and-Wait - ARQ*, *Go-Back-N - ARQ*, *Selective Repeat - ARQ* and the Transmission Control Protocol (TCP) are used. To derive an analysis of the performance of the protocols, a trace-based simulation is executed using the OMNeT++ software, so that the errors due to fading in the optical channel are simulated using previous measurements of real FSO links.

Another different approach to simulating an optical channel is through the use of probabilistic models. In [5] can be seen an analysis of a retransmission protocol, in which a model with the Gamma-Gamma distribution is used to analyze the physical layer.

In the articles mentioned in the previous paragraphs, a teletraffic analysis of the data transmission between a CubeSat satellite and a ground station is not carried out, in such a way that a protocol can be evaluated based on its performance measurements. For that reason, it was proposed a project to develop a discrete event simulator to represent with simple models (optical channel and satellite orbit) the FSO link between a CubeSat and a ground station, in order to evaluate the behavior of data link protocols and flow control strategies on that environment that is being simulated. The Selective Repeat ARQ (SR-ARQ) was the first selected protocol to be evaluated because it is more efficient than Stop-and-Wait and Go-Back-N. Some improvements are being proposed in the standard SR-ARQ to adapt it to optical satellite links taking into account the change in satellite distance during a line-of-sight period and the impairments of the optical channel. To evaluate the proposed improvements and compare them with the standard SR-ARQ, it is proposed to use the *throughput* as a performance parameter, and use traffic generators to generate the traffic necessary to estimate that parameter.

This paper presents the results of research on the types of traffic generators that exist and the performance parameters that can be measured with them. This is only one of the issues necessary in order to do the simulation proposed in the main project.

In section 2 an explanation of the traffic generators and the performance parameters used to do the evaluation of the SR-ARQ protocol is given. In section 3 the context and considerations about the system being simulated are explained and the results of the research on how to adapt and apply them to perform a teletraffic analysis are shown. Finally, in section 4 the results are summarized and the conclusion of the research is presented.

2. Materials and methods

2.1. Performance parameters

First, it is necessary to understand the performance parameters used to evaluate the effectiveness and efficiency of data transmission over digital communication systems such as data link protocols or networks:

- *Protocol overhead*: It is all data that is not part of the payload but is nonetheless transmitted in a communications system to effectively support the operation of the protocol. The overhead includes preambles, frame headers, acknowledgment packets, and erroneous packets that had to be retransmitted.
- *Throughput*: It is a performance parameter that indicates the rate at which a communications system delivers to a recipient the data that a sender transmits (subtracting overhead introduced by the system), given a certain amount of time. It can be measured in bits per second (bps) or packets per second (pps), depending on the communication system or the purposes for which it is being measured. When throughput is calculated or measured at the application layer, it is called *goodput.*
  Different types of throughput can be defined [7] [8]:
  - *Maximum theoretical throughput*: It is the maximum throughput that a specific communications system can achieve assuming the best conditions, i.e. when the system is assumed to add the minimum overhead (which must be subtracted from the payload to calculate throughput).
  - *Maximum asymptotic throughput*: It is the maximum throughput calculated or estimated through simulation that a communication system can achieve when the load on its input approaches infinity, either because of the size of the message sent to a data link (system) approaches infinity or because the number of data sources sending messages on a network (system) is very high, no matter how much overhead the system adds. Some authors name the maximum asymptotic throughput as maximum sustainable throughput [9] [10].
    A greedy source can be used to estimate asymptotic throughput.
  - *Instantaneous throughput*: It is the measured throughput at the limit when the time approaches zero. On the other hand, *peak measured throughput* is the maximum instantaneous throughput measured in a certain period.
  - *Average throughput*: It is the instantaneous throughput averaged or integrated over a long time. When the load on the system input is very high, it approaches the maximum asymptotic throughput.
- *Channel utilization*: It is the percentage of the transmission data rate established on a link that is used in the transmission of both payload and overhead data.
- *Channel efficiency*: It is the percentage of the transmission rate established on a link that is used in the transmission of payload data. It is the ratio of the throughput divided by the transmission data rate established in the channel.

2.2. Stochastic Processes and Probability Distributions

Traffic generators use some stochastic processes and probability distributions in order to generate data. Following, some of those stochastic processes and probability distributions are defined:

- *Bernoulli Process:* It is a discrete-time stochastic process consisting of a finite or infinite sequence of independent discrete random variables X_i, such that:
  - for each i, the value of X_i is equal to 1 or 0 (success or failure).
  - for all value of i, the probability p that $X_i = 1$ (success) is the same.

Thus, a Bernoulli process is a sequence of independent Bernoulli trials with identical distribution, which can result in success with probability p and failure with probability $q = 1 - p$.
Independence in trials implies that the process has no memory, that is, past results do not provide information about future results.

- *Binomial distribution:* The binomial probability distribution, characterized by the random variable X, represents the probability that x successes are obtained in a Bernoulli Process with n trials, given a probability of success p. It is represented as follows:

$$b_X(x; n, p) = \binom{n}{x} p^x (1-p)^{n-x} \tag{1}$$

Where $E[X] = np$ is the average, and $Var[X] = np(1-p)$ is the variance.

- *Poisson Process:* It is a continuous-time stochastic process that consists of a succession of events in a given time interval, in such a way that:
  - The number of events that occur in a time interval is independent of the number of events that occur in any other disjoint time interval. Therefore, the Poisson Process has no memory.
  - The probability of a single event occurring in a very short time interval is almost proportional (with an insignificant difference) to the size of the interval, and does not depend on the number of events that occur outside the interval.
  - The probability of more than one event occurring in such a short time interval is insignificant.
- *Poisson distribution:* The probability distribution of the discrete Poisson random variable Y, which represents the probability of a given number of events occurring in a continuous time interval t, given a rate of events per unit time α, is defined as [11]:

$$p_Y(y; \kappa) = \frac{e^{-\kappa}(\kappa)^y}{y!} \tag{2}$$

Where $E[Y] = Var[Y] = \kappa = \alpha t > 0$ is the average and the variance.
If for a binomial probability distribution $b_X(x; n, p)$, it holds that $n \to \infty$ and $p \to 0$, such that $np \to \kappa > 0$, then $b_X(x; n, p) \to p_Y(y; \kappa)$ with $y = x$. That is, in any Bernoulli Process in which numerous trials are performed $(n \to \infty)$ and the probability of success is small $(p \to 0)$, an approximation to the probability of the number of successes x can be done by using the Poisson distribution, since, in such a case, $b_X(x; n, p) \approx p_Y(y; \kappa)$ with $y = x$.

- *Exponential distribution:* A continuous random variable Z has an exponential distribution with parameter $\lambda > 0$ if its density function is given by [11]:

$$f_Z(z; \lambda) = \begin{cases} \lambda e^{-\lambda z} & z \geq 0 \\ 0 & \text{otherwise} \end{cases} \tag{3}$$

Where $\mu = E[Z] = \frac{1}{\lambda}$ is the average, and $Var[Z] = \frac{1}{\lambda^2}$ is the variance.
Assuming that the number of events that occur in a time interval t follows a Poisson distribution with parameter $\kappa = \alpha t$, with α being the rate of occurrence of events per unit of time, then the distribution of the time elapsed between the occurrence of two successive events (z) in said interval follows an exponential distribution with parameter $\lambda = \alpha$.

- *Weibull distribution:* A random variable V has a Weibull distribution with parameters $\alpha > 0$ and $\beta > 0$ when its probability density function is given by:

$$w_V(v;\alpha,\beta) = \begin{cases} \frac{\alpha}{\beta}\left(\frac{v}{\beta}\right)^{\alpha-1} e^{-(v/\beta)^{\alpha}} & v \geq 0 \\ 0 & \text{otherwise} \end{cases} \quad (4)$$

Where $\gamma = E[V] = \beta\Gamma\left(1+\frac{1}{\alpha}\right)$ is the average, and $Var[V] = \beta^2\left\{\Gamma\left(1+\frac{2}{\alpha}\right) - \left[\Gamma\left(1+\frac{1}{\alpha}\right)\right]^2\right\}$ is the variance, and Γ is the gamma function.
When $\alpha = 1$, the probability density function reduces to the exponential distribution.

2.3. Discrete event simulations
A discrete event simulation is a process through which a Discrete Event System (DES) model, such as Markov chains, is numerically evaluated, and the data obtained in this process are used to estimate quantities of interest [12].

An analytical solution for a DES is generally complex, for this reason, simulation is very useful.

2.4. Traffic generators
Below, the traffic generators characterized in the project are defined:

- *ON-OFF traffic generator:* consists of a finite state machine with two states: ON and OFF. In the ON state, the generator emits a burst of separate packets at regular time intervals, Δt. In the OFF state, the generator remains silent and does not emit packets.
  There are different types of ON-OFF traffic generators, depending on the probability distribution used to characterize the duration of the ON and OFF periods [13] [14]. Two types are mentioned here:
  - Exponential ON-OFF traffic generator.

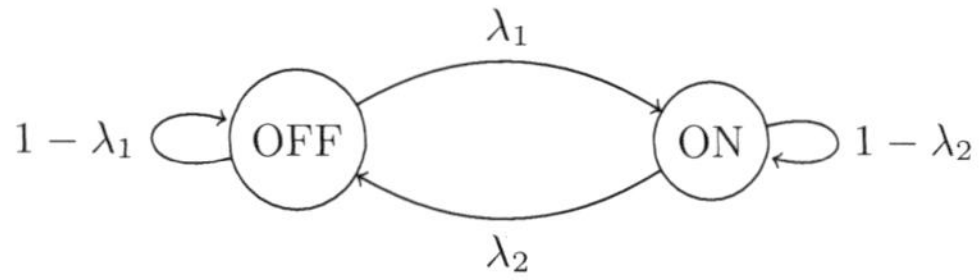

Figure 1. Finite Markov Chain for the exponential ON-OFF traffic generator.

    The exponential distribution is used to characterize the duration of the ON and OFF periods. Considering that $\mu = \frac{1}{\lambda}$ is the mean of the exponential probability distribution, then:
    * $\mu_{on} = \frac{1}{\lambda_1}$ is the average time in the ON state.
    * $\mu_{off} = \frac{1}{\lambda_2}$ is the average time in the OFF state.
    * The probability of being in the OFF state is $P_{off} = \frac{\lambda_2}{\lambda_1+\lambda_2}$, and the probability of being in the ON state is $P_{on} = \frac{\lambda_1}{\lambda_1+\lambda_2}$.
  - Weibull ON-OFF traffic generator.
    The Weibull distribution is used to characterize the duration of the ON and OFF periods. Considering that $\gamma = \beta\Gamma\left(1+\frac{1}{\alpha}\right)$ is the mean of the Weibull probability distribution, then:
    * γ_1 is the average time in the ON state.
    * γ_2 is the average time in the OFF state.

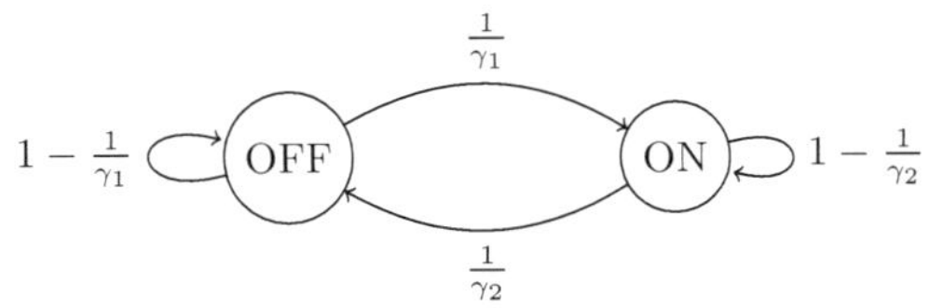

Figure 2. Finite Markov Chain for the Weibull ON-OFF traffic generator.

The ON-OFF traffic generator model is generally used to simulate the behavior of heavy-tailed traffic in a data network environment. To do this, several ON-OFF sources are usually multiplexed. In these cases, some other heavy-tailed distributions are used to characterize the ON and OFF periods.
Since we want to simulate the traffic in a satellite link where two devices interact, the satellite and the ground station, it was hypothesized that the traffic could be simulated by a single ON-OFF source.
Having simulated the behavior of the traffic in the satellite link by means of an ON-OFF generator, it is possible to estimate the average throughput.

- *Greedy traffic generator:* it is a traffic generator that broadcasts data at the highest possible rate and as soon as there is an opportunity to do so.
This type of generator is helpful in estimating asymptotic throughput through simulation, regardless of whether or not the generator behaves like a real data source. Therefore, the maximum capacity of a system to transfer data from a source to a destination can be dimensioned.

2.5. Traffic consumer
The traffic consumer is the destination of the messages sent by the traffic generator. There are two strategies by which the traffic consumer can get data:

- *Push strategy*: the SR-ARQ receiver module pushes the segments in order to the traffic consumer, as soon as it is possible.
- *Pull strategy*: the traffic consumer request the SR-ARQ receiver module for segments independently, according to a certain probability distribution, by sending requests via the input priority queue of the SR-ARQ receiver module.

3. Results and discussion
First, subsection 3.1 gives context about the system that is being simulated in the main project, in order to locate where are traffic generators located and why they were the selected approach to estimate performance parameters. The remaining subsections explain the considerations and adaptation to the simulator of the traffic generators and the traffic consumer defined in the previous section.

3.1. Design considerations
The primary objective was to estimate throughput on the application layer (goodput) that an FSO channel with a selective repeat ARQ protocol implemented in the data link layer could achieve. To estimate that, it was considered the following:

- The system could not be implemented on a real link, as there were several limitations.

- If the system was mathematically modeled, it would be required to use a mix of various models and parameters, so that it would be a highly complex model. For that reason, it was decided to use a simulation strategy. To simulate the packet transmission following the Selective Repeat protocol, it was built a discrete event simulator.
- The specific application of the CubeSat satellite was not delimited. For that reason, a specific data source and an application layer protocol were not characterized. The application layer protocol defines characteristics like the protocol of request for data.

For the above reasons, it was decided to use data traffic generators in a discrete event simulator to represent the behavior of the selective repeat ARQ protocol in an FSO link.

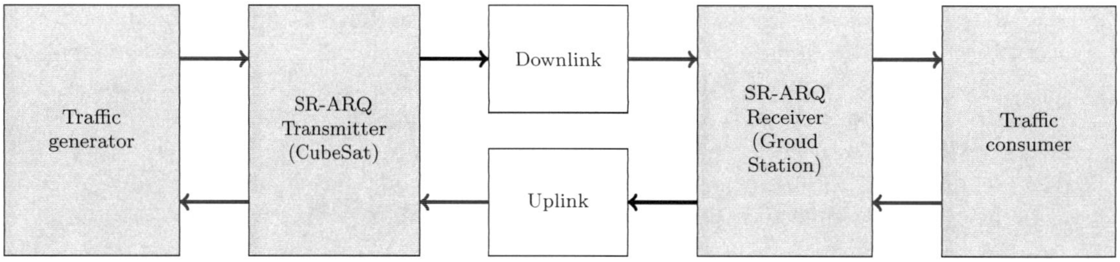

Figure 3. Simulator's modules

The simulator is represented in figure 3. It is important to point out the following:

- Modules colored blue run on different execution threads. Each one has a clock that is updated as events occur.
- Communication between modules running on separate threads is done via priority queues. That is, each module has an input and output priority queue. The messages that arrive at the modules colored blue are considered events whose priority is given by the time in which these events occur. In a module, the event (message) with the highest priority (shortest time) is attended to first.
- The link modules simulate the satellite FSO channel. They are not executed in independent threads, they are just elements that act according to how they are used by the other modules.
- The traffic generator module generates segments of data and puts them in the input priority queue of the SR-ARQ transmitter module as events whose timestamps are the completion time in which each segment was completely generated. When the SR-ARQ Transmitter module gets an event from its priority queue whose origin is the traffic generator, then it unpackages the segment and attempts to encapsulate it in a frame and store it in the transmission/retransmission buffer.

3.2. ON-OFF traffic generator implementation
In the literature, it is common to see that exponential ON-OFF traffic generators are used when modeling voice traffic over a packet-switched network. A relationship with the Poisson arrival process, used in telephony to represent independent call requests, can be observed. For that reason, in the satellite data link being simulated, the exponential distribution in the state change from OFF to ON could represent that the file download requests occur independently. In the same way, by using an exponential distribution for the state change from ON to OFF, it could represent the independence of the lengths of the files if it follows an exponential distribution. However, it is necessary to make clarifications.

The problem using an exponential probability distribution is that, following the assumption above, the change of state from ON to OFF is more probably to occur in a short time interval, and therefore the ON state is more probably to have a very short duration. In this case, it would mean that the amount of small-length files is greater than the quantity of large-length files, which is not very useful for representing high data transmission demands.

On the other hand, since the probabilities of changing states from OFF to ON and from ON to OFF each follow an exponential distribution with different parameters, the rate of changes from one state to another, no matter what the change is, follows a two-parameter hypoexponential distribution (the convolution of two random variables with an exponential distribution) [15]. That is, until the moment in which the system falls in one state, the time for the change to the other state is calculated (a realization of the random experiment with exponential distribution is carried out). Time accumulates. Thus, changes of state from OFF to ON, which could be interpreted as file download request arrivals, do not occur as a Poisson arrival process.

In [16], traffic on a personal computer was statistically measured. The periods in which there is traffic or not were characterized by measuring the load (instantaneous throughput) and comparing it with a certain decision threshold. Based on the durations of the periods with traffic and without traffic, two graphs were constructed, which represent the rate of times each of the states occurs as a function of its duration. Subsequently, the two graphs were approximated by using an ON-OFF traffic generator, testing with different probability distributions: exponential, Weibull, Gamma, geometric, and Poisson. The result was that the graphs constructed by means of the ON-OFF generator are more similar when a Weibull distribution is chosen. If the graph of the Weibull distribution is observed, it can be noticed that, unlike the exponential distribution, the probability that the duration in a state tends to zero is low.

Regardless of the probability distribution used in the ON-OFF traffic generator, in future work, measurement of the traffic in a real satellite should be made to determine which probability distribution and with what parameters is more similar to the duration of the ON and OFF states. Both the exponential distribution and the Weibull distribution were implemented in the simulator.

Figure 4 shows the implementation of the exponential ON-OFF traffic generator. The program was made with Python and Numpy. Since what matters is the duration of the ON and OFF states, the implementation is simple:

(i) The traffic generator starts with a clock initialized to zero and is in the OFF state.

(ii) Immediately, a realization of the pseudo-random generator of the OFF to ON transition time is obtained. This time corresponds to the duration of the OFF state.

(iii) The generator changes state to ON and the time obtained in the pseudo-random generator is added to the clock.

(iv) Upon switching to the ON state, a realization of the pseudo-random generator of the ON to OFF transition time is immediately obtained. This time corresponds to the duration of the ON state.

(v) In the ON state, instead of generating fixed-length packets at a constant inter-arrival time, as defined in the standard ON-OFF generator, it was decided to generate a continuous bit stream whose length is determined by the product of the duration of the ON state and the bit rate demanded by the data source. Said bit stream is generated following a Bernoulli Process with an equal probability of success and failure.

(vi) The clock at its current time represents the start time at which the bit stream began to be generated. Once the bit stream is generated, the time obtained in the pseudo-random generator from the ON to OFF change is added to the clock. This new time corresponds to the end time of the generation of the bit stream.

(vii) The stream of generated bits is segmented (according to the maximum size allowed by the SR-ARQ protocol defined in the transmitter) and the completion time of the generation of each segment is calculated. Each segment is sent to the input priority queue of the SR-ARQ transmitter module, whose priority is given by the segment generation completion time.

(viii) Steps (ii) to (vii) are repeated until the clock reaches the finalization time.

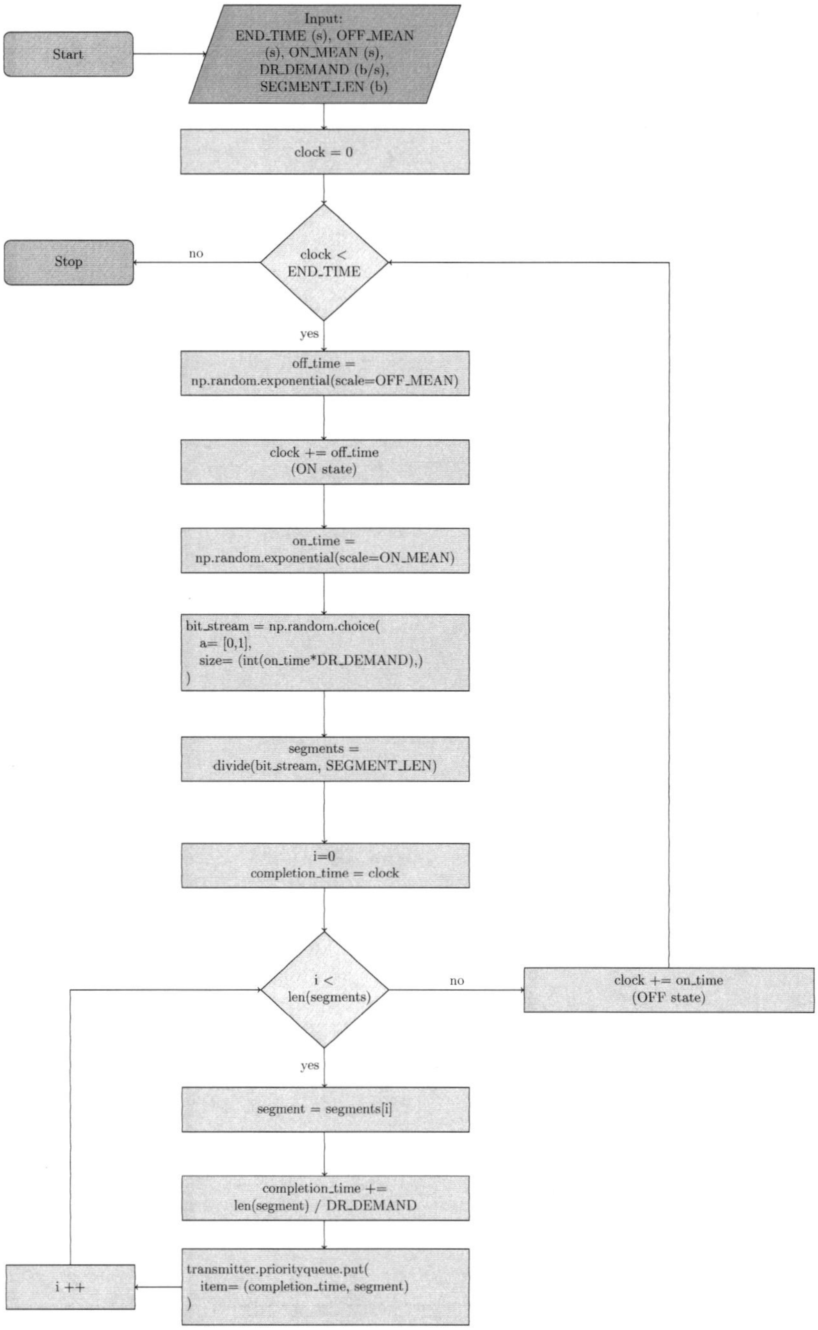

Figure 4. Exponential ON-OFF traffic generator implementation

The Weibull ON-OFF traffic generator implementation is similar to the exponential ON-OFF traffic generator implementation

The transmitter module in due time (represented by its clock) will or will not insert the received segment (after encapsulation) in its transmission/retransmission buffer, depending on the free spaces it has. In the case that the insertion is not made, there are three options:

- Give the segments up for lost.
- Retry sending rejected segments after a certain time. To do this, it is necessary to specify the retry time, the maximum number of retries (if it exists), and the limit on the number of segments that can be stored in the traffic generator to retry sending them later.
- Use a strategy where the sender notifies the generator when buffer space is freed. In this case, the scheme explained above would change in that the generator would not send the segments to the transmitter's input priority queue, but would store them to send until the transmitter notifies the number of free spaces in the buffer and the time of the transmitter's clock in which these free spaces are found.

The first and third options were developed.

The implementation of the ON-OFF traffic generator in the simulator is done in order to represent different traffic loads in the selective repeat ARQ protocol to measure the instantaneous and average throughput and not just the limiting case given by the greedy traffic generator (asymptotic throughput).

In the satellite context, throughput estimates are only valid when there is line-of-sight, for this reason, it is emphasized that teletraffic is carried out in the line-of-sight period. On larger time scales, during an orbital period, the traffic cannot be represented solely with an ON-OFF generator and models will be required that not only represent the throughput as a function of time, but also of space.

3.3. Greedy Source implementation

This traffic generator allows estimating the maximum capacity of the link, in such a way that an estimate of the *asymptotic throughput* can be obtained.

To generate the largest possible data load, it is enough for the traffic generator to generate enough segments and send them to the SR-ARQ transmitter to keep the transmission/retransmission buffer full. To accomplish this, the SR-ARQ transmitter notifies the traffic generator of each state change using the aforementioned communication mechanism. In this way, the traffic generator knows how many buffer spaces are free and, consequently, can produce enough segments to send to the transmitter and keep the buffer full.

3.4. Traffic consumer implementation

For simplicity, assuming that in the ground station there is no bottleneck, the push strategy was developed. With this strategy, it is possible to evaluate the asymptotic throughput in the case that the greedy traffic generator produces traffic.

4. Summary and conclussion

Two types of traffic generators were studied, the ON-OFF traffic generator and the greedy traffic generator. By using the ON-OFF traffic generator, different probability distributions can be used. In the table 1 it is summarized the traffic generators presented in this paper.

Traffic generators are useful to perform a teletraffic analysis on real or simulated data links or networks. In this case, a discrete event simulator of an FSO link between a CubeSat and a ground station was designed. The objective was to determine the throughput of an SR-ARQ protocol implemented in that link being simulated.

Journal of Physics: Conference Series **2475** (2023) 012003 doi:10.1088/1742-6596/2475/1/012003

Table 1. Traffic generators

Feature	**Exponential ON-OFF generator**	**Weibull ON-OFF generator**	**Greedy source**
Bit stream generated by	Bernoulli Process with equal probability of success and failure.	Bernoulli Process with equal probability of success and failure.	Bernoulli Process with equal probability of success and failure.
Bitstream structure	Variable length bitstreams, that are split into fixed length segments. Small-length bitstreams are more probable.	Variable length bitstreams, that are split into fixed length segments. Not necessarily small-length bitstreams are more probable.	Contiguous bitstream with virtually infinite length, that is split into fixed-length segments.
Estimated throughput	Average	Average	Asymptotic

Thanks to those tools it will be able to evaluate and propose different improvements to the data link protocols and flow control strategies in order to implement them in a CubeSat to ground station FSO link, in the future. The main project of the simulator is still a work in progress.

References

[1] Davoli F, Kourogiorgas C, Marchese M, Panagopoulos A and Patrone F 2019 *Int. J. of Satellite Communications and Networking* **37** 343–359

[2] The Consultative Committee for Space Data Systems 2022 What is ccsds? URL https://public.ccsds.org/default.aspx

[3] Le H 2022 Link-Layer Retransmission-based Error-Control Protocols in FSO Communications: A Survey

[4] Jahid A, Alsharif M H and Hall T J 2022 *J. of Network and Computer Applications* 103311

[5] Le H D, Mai V V, Nguyen C T and Pham A T 2019 *J. of Optical Communications and Networking* **11** 151–163

[6] Epple B 2007 Performance optimization of free-space optical communication protocols based on results from fso demonstrations *Free-Space Laser Communications VII* vol 6709 (SPIE) pp 335–346

[7] Ueberhuber C W 1997 *Numerical computation 1: methods, software, and analysis* 1st ed (Berlin: Springer) ISBN 3-642-59118-3

[8] Chou C Y, Chang H Y, Wang S T and Tcheng S C 2006 Modeling message-passing overhead on nchc formosa pc cluster *Advances in Grid and Pervasive Computing* ed Chung Y C and Moreira J E (Berlin, Heidelberg: Springer Berlin Heidelberg) pp 299–307 ISBN 978-3-540-33810-9

[9] Imai S, Patterson S and Varela C A 2017 Maximum sustainable throughput prediction for data stream processing over public clouds *2017 17th IEEE/ACM International Symposium on Cluster, Cloud and Grid Computing (CCGRID)* pp 504–513

[10] Ohlinger M, Sharkey K and Cai S 2021 Measuring maximum sustainable engine throughput URL https://learn.microsoft.com/en-us/biztalk/core/measuring-maximum-sustainable-engine-throughput

[11] Devore J 2005 *Probabilidad y estadstica para ingeniera y ciencias* (Mexico City: International Thomson Editores)

[12] Cassandras C G and Lafortune S 2008 *Introduction to discrete event systems* (Springer)

[13] Parra León A, Piedrahita E M and Salcedo O 2011 *Tecnura* **15** 129–147

[14] Garcia A, Hackbarth K, Brand A and Lehnert R 2002 *WSEAS Transactions on Communications* **1** 59–65

[15] Staake T R 2002 *IP Traffic Statistics: A Markovian Approach* Ph.D. thesis Citeseer

[16] Gustafsson E 2020 *Synthetic Generation of Realistic Network Traffic* Master's thesis Department of Computer and Information Science, Linkping University

Journal of Physics: Conference Series **2475** (2023) 012004 doi:10.1088/1742-6596/2475/1/012004

Quartz-based Passive Signal Sorter as Radiofrequency Receiver

I Medina[1], D Martínez[2], D E Suárez[2] and O Pineda[2]
[1]Instituto Politécnico Nacional, Centro de Desarrollo Aeroespacial, Mexico City, Mexico
[2]Instituto Politécnico Nacional, Escuela Superior de Ingeniería Mecánica y Eléctrica Unitad Zacatenco, Mexico City, Mexico

E-mail: imedinas@ipn.mx

Abstract. In this research, we developed a quartz-based passive signal sorter as a radiofrequency receiver, which wirelessly generates voltage from frequency, particularly NFC (13.56MHz), with high efficiency and energy use. Through the electromagnetic field, the passive signal sorter takes advantage of the environment's frequency to make the quartz oscillate and generate voltage as a result of the piezoelectric effect, along with diodes and capacitors to rectify the electric signal. Among its possible uses, it can be applied to a CUBESAT, since it currently implements solar supply through panels and stores this energy in batteries, because the normal environment limits several types of supply. The CUBESAT spends great periods of time without using direct sunlight, and the farther it is from it, the longer it takes to fill the batteries, and in some particular cases, to reduce the subsystems' charge. It is suggested that the quartz-based signal sorter works as a reserve source that supplies these batteries by getting energy from the Earth's electromagnetic fields, preventing as much as possible the lack of energy. The circuit's development leads to a supply proposal for a CUBESAT system by proving it is economically and volumetrically viable and that it is attractive not only because of the aforementioned benefits but also for being highly efficient and friendly to the environment. It is a great opportunity to implement it in space, industrial, corporate, and household sectors.

1. Introduction

From the first electronic circuits that used quartz, there has been a lot of research regarding its piezoelectricity, conductivity, resistance, and frequencies. The most usual frequencies to which they have been submitted vary from Kilo Hertz (kHz) to Giga Hertz (GHz), resulting in different oscillatory and conductive behaviors. This wide range has been proven in several studies on quartz sensitivity, as well as on their electromagnetic fields [1][10][9][12].

Quartz is the most abundant mineral on Earth; belongs to the silicate group (SIO2), and both its hardness (7 on the Mohs Scale) and its piezoelectric properties are some of its most valued qualities in different industries [7].

For example, quartz has been used on oscillating circuits, because they take advantage of the mechanical resonance of piezoelectric material to generate an electric signal with a specific operating frequency. This effect has several uses, like in watches, to provide constant signaling in electronic circuits [11].

Besides, piezoelectricity is considered a fundamental characteristic in quartz crystals, since they are highly sensitive materials to both frequencies present in nature and other ones generated through means such as antennas or electronic devices [4].

The oscillating quartz crystal is exemplified in Figure 1, which shows its content on an insulating metal packing that blocks the entrance of imperfections or pollutants and helps to it have a correct solid assembly for the crystal. The lower pegs or pins allow to connect the crystal to a Protoboard or PCB to weld the components. These pins are directly connected to the metal-protected support strut and have two electrodes around the crystal to transmit the necessary frequencies for the crystal to work, in this case, transmitting 13.56 MHz [8].

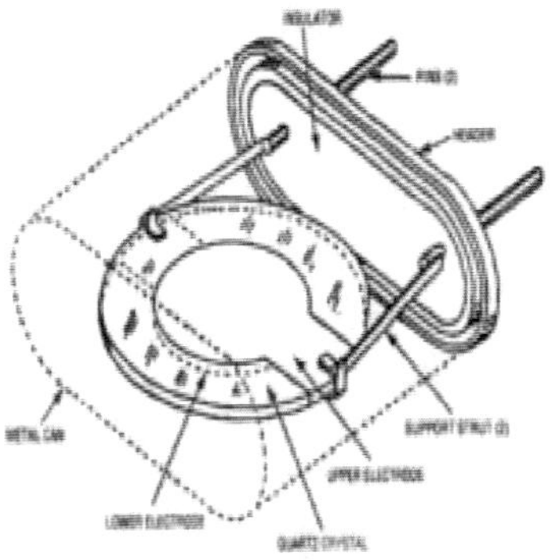

Figure 1. 2D Representation of Quartz Oscillator 13.56 MHz [8].

Regarding the life cycle of oscillating quartz crystals, several factors are taken into account. The first one is proper maintenance: they have to be externally cleaned up with isopropyl alcohol in order to remove any impurities from the component and ensure its correct operation.

Amongst the most common alterations, there are internal fractures, extreme temperature exposure, and crystal aging, which can result in interferences and frequency glitches while oscillating.

Due to these factors, there isn't an accurate or exact life cycle for these elements; however, if they are maintained in controlled conditions, they can last up to 20 years. The most recommendable thing to do is to replace them periodically (every five years) since the material is actually quite affordable [13].

The dimensions of the crystal used are shown in Figure 2, along with its picture as a graphic description, and a 2D diagram expressed in millimeters.

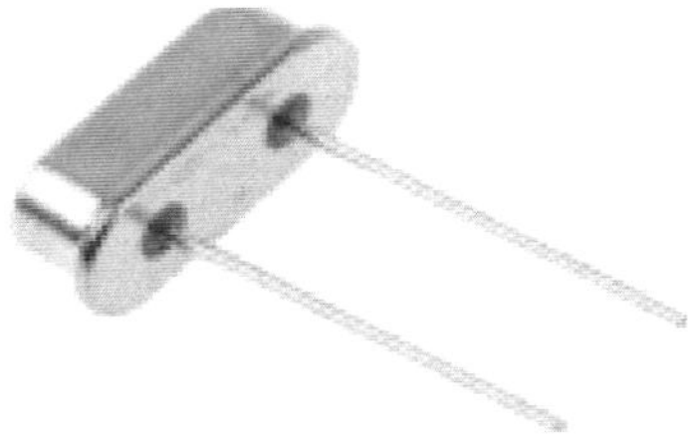

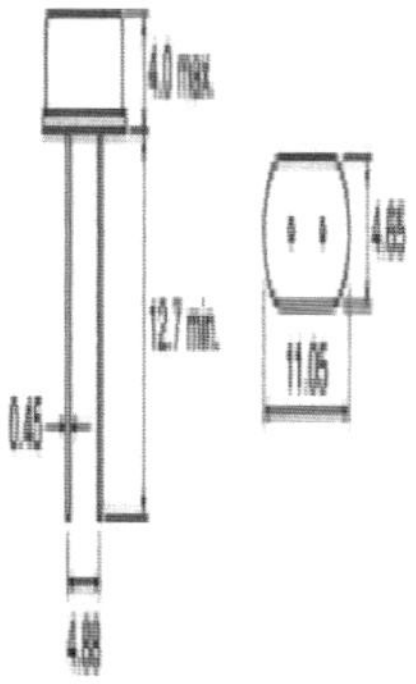

Figure 2. 2D Representation of the Quartz Oscillator 13.56 MHz [3].

In order for the sorter to work as a reserve source, we require electromagnetic fields, but there currently exists a very safe technology, NFC (Near Field Communication). NFC is a wireless technology regulated by ISO standards, and its main function is to communicate or connect two devices through low frequencies, exchanging information between them.

The technology operates on a 13.56 MHZ frequency, and its operation is based on the electromagnetic induction principle [2].

It has two operation modes [5]:

- Passive mode: An active device with its own supply source generates an electromagnetic signal, and a passive element without a supply source uses a part of the electromagnetic field energy for supply.
- Active mode: Both devices must have their own supply source to be able to generate their own electromagnetic field in order to transmit and receive data.

The sorter's use is passive since it doesn't need its own supply source to send signals. It just requires receiving the frequency to make the oscillatory effect.

2. Rectenna

The transmission and reception of the frequency signal were implemented through the ANFCA-101- 2515-A02 antenna that typically works with NFC signals at 13.56 MHz, which is very useful for the application and supply of the quartz crystal at the same operating frequency [6].

The antenna requires several specifications to have an optimal operation and get the best possible signal reception, among them, a 1-centimeter distance to receive the greatest frequency, considering that the closer it is from the 13.56 MHz signal source, the better. The ideal location is pointing towards the signal source on a horizontal axis, with the receiving surface fully aligned to the transmitting device.

If the antenna is placed near a metallic piece or coils that generate magnetic fields, this can cause loss or interferences on the signal received (it is best to avoid these objects within a radius of 20 centimeters). Using devices and/or antennas of the same brand and technology can increase its effectiveness [6].

Table 1 below shows a comparison among different wireless technologies, which can also be implemented in the prototype, as well as their pros and cons when assessing them with different parameters.

Table 1. Comparison among technologies based on [14].

Features	nfc	bluetooth	wifi	zigbee	irda
Communication set-up time	Less than 0.1 seg	6 s	4 s	30 ms	0.5 s
Transmission speed	424 kbps to 848 kbps	2.1 Mbps (3.0 version supports 24 Mbps)	1000 Mbps	250 kbps	115 kbps
Range	20 cm	Up to 10 m	100 m	70 m	1 m
Energy use	Low	High	High	Medium	Medium
Cost	Low	Medium	High	High	Low
Safety	High	High with encryption	Medium	128-bit AES encryption	Medium

In this chart, we can see that NFC technology has greater advantages, especially in the communication set-up time and energy use, besides having high safety. Although its transmission speed is not as high compared to WIFI or Bluetooth, it is enough for the prototype.

The frequency interval where the passive signal sorter works is between 13.5675 and 13.569 MHz, which is still within NFC parameters [12].

According to Figure 3, the voltage obtained not only depends on frequency but also on absolute power (dB); the lesser the power (-35 dBm), the lesser the energy transmitted ($\approx .5 volts$), and the higher the power (-20dBm), the higher the energy transmitted (5 volts) [12].

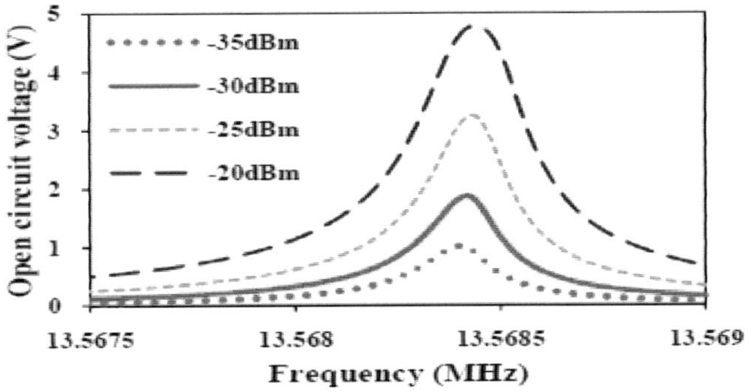

Figure 3. Graph of voltage in frequency domain obtained from paper "A New Family of Passive Wireless RF Harvesters based on R-C-Quartz Oscillators" [12].

The prices shown in Table 2 were gathered between September and October of 2022, which increased due to the worldwide pandemic and the scarcity of electronic materials in the country.

Table 2. Cost of materials for the prototype. Own research.

Materials	Price USD
Resistor 47Ω	0.255
Resistor 10kΩ	0.255
2 diodes (1n5240b)	1.53
Capacitor 2.7 pF	0.255
Capacitor 100 pF	0.255
Oscillating Quartz at 13.56 MHz	1.275
2 antennas for 13.56 MHz signal (NFC)	15.3
Protoboard	5.1
2 meters of cable 22-AWG caliber	1.02
TOTAL	25.245

Given the aforementioned situation, some materials to build the prototype didn't have an optimal distribution, such as the HSMS-286C rectifier; therefore, it was replaced by a pair of Schottky diodes. Also, the finest and most accurate oscillating crystal was replaced by a more commercial one.

The oscillation/resonance frequency of the quartz chosen was 13.56 MHz, mainly because we wanted to receive easily available (commercial) low signals that were already present in the environment, such as radio or television, as well as frequencies that had higher resistance to interferences and operating distances, compatible even with the new NFC technologies, giving as an alternative for greater distances WIFI signal frequencies at 2.4 GHz.

The circuit receives a frequency through the receiving antenna, and this signal excites the oscillating quartz crystal, since its oscillating frequency is 13.56 MHz, and starts to spread current through the circuit. With the two Schottky diodes, it is made the mean wave rectification to be able to get positive peaks of the full wave, obtaining the output voltage in direct current.

The capacitors store the energy so that when the circuit commutes, it keeps some of it in order to always generate a voltage on the output.

The 13.56 MHz NFC signal is generated by an 'ANFCA- 101-2515-A02' NFC antenna, which is supplied with a voltage of 0.5 to 1 Volt, and has an electronic one that allows it to act as a transmitting or receiving antenna; although, for practicality, it only transmits electromagnetic field at the required frequency, which can be received by the receiving antenna. For the simulation of the circuit shown in Figure 4, it was implemented the Proteus application, with the following components:

(i) ✓2 resistors (50 and 10k Ohms)
(ii) ✓2 diodes (1N5240B)
(iii) 2✓capacitors (2.7pF and 100pF)
(iv) ✓Quartz oscillator at 13.56 MHz
(v) ✓Signal Generator of 13.56 MHz
(vi) ✓CD Voltmeter
(vii) ✓Ground

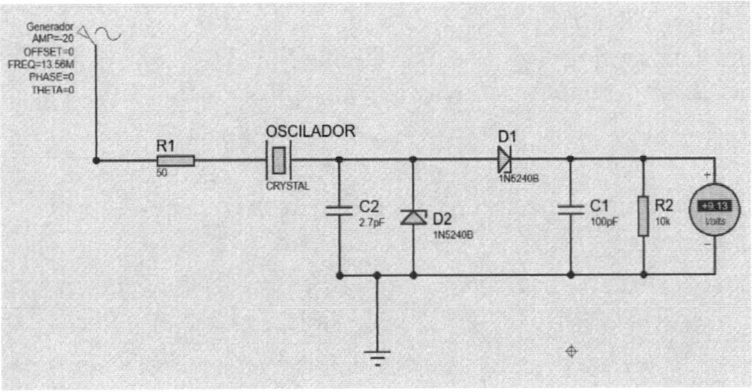

Figure 4. The circuit in the simulation of the RF receiving antenna.

As can be seen in Figure 5, the circuit of the radiofrequency receiving antenna is small, which favors its use on CUBESAT.

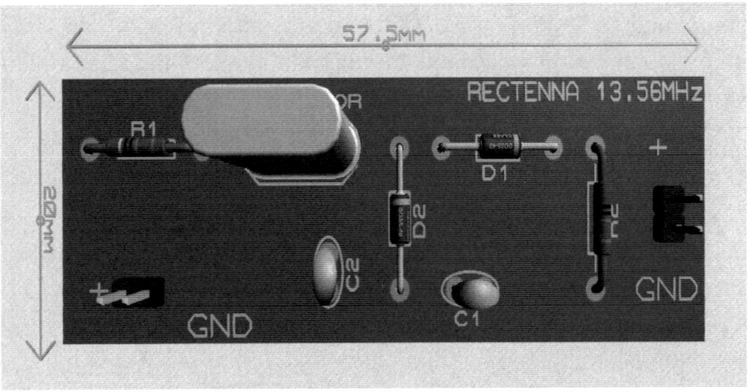

Figure 5. 3D representation of the circuit, measured in millimeters.

The maximum output signal was 9.13 V, as long as the quartz receives the frequency to oscilate (13.56 MHz) the circuits output will constant, however, if the frequency is modified out of crest of graphic or the components are changed, the voltage will be affected (from mV at V); therefore, it is deemed a successful result, considering the simulation on the software and the best value of each element.

3. Conclusions

Quartz and its piezoelectric property create enough tension to supply a battery in a CUBESAT, being extremely important for the accuracy of both the frequency and decibels for the communication between crystals, which is controlled wirelessly through electromagnetic fields. The functionality of the passive radiofrequency receiving antenna was proven in the space sector: it was strong and resistant to environmental noise and interference, and its small size proved to be a great advantage.

One of the quartz's main characteristics is its resonance capacity which, along with its piezoelectricity, allows it to create voltage through frequency reception, due to its high sensitivity. The most commercial quartzes are the low-frequency ones; however, for the development of the quartz-based passive signal sorter as a radiofrequency receiver, the optimal one is of 13.5685MHz.

We want to point out and highlight that the circuit's high efficiency of 1,000% surpasses most of the wireless technologies at a relatively low cost, supplying with 1 Volt the transmitting antenna and obtaining $\approx 10Volts$.

Acknowledgments This project has been partially supported by project 20221182, as well as EDI grant, all provided by Secretaría de Investigación y Posgrado, Instituto Politécnico Nacional.

References

[1] A. Bell. *Enciclopedia Británica. Reino Unido, Estados Unidos: Encyclopedia Britannica.* 2010.

[2] M. F. Carignano. *NFC (Near Field Communication).* 2009. URL: https://rdu.iua.edu.ar/bitstream/123456789/462/1/TFI_ESE_Maria_Fernanda_Carignano%20.pdf.

[3] *Cristal de cuarzo 16MHZ (pequeño) ¿ cristal / oscilador / filtro ¿ componentes electronicos ¿ cristal de cuarzo ¿ pequeño.* www.planetaelectronico.com. URL: https://www.planetaelectronico.com/cristal-de-cuarzo-16mhz-pequeno-p-19901.html (visited on 01/15/2023).

[4] Un Cristal Piezoeléctrico-E et al. *Resonancia en Laboratorio 4-Dpto. de Física -FCEyN-UBA-1996.* URL: http://users.df.uba.ar/sgil/web_fisicarecreativa/informes/infor_ondas/piezoelectr96.pdf.

[5] Bueno Delgado et al. *La tecnología NFC y sus aplicaciones en un entorno universitario.* 2011. URL: https://repositorio.upct.es/bitstream/handle/10317/2494/2.1.pdf?sequence=1&isAllowed=%20y.

[6] . *FLEXIBLE NFC ANTENNA WITH FERRITE ANFCA-101-2515-A02.* 2018. URL: https://www.mouser.mx/ProductDetail/ABRACON/ANFCA-101-2515-%20A02?qs=wUXugUrL1qzYcgV1R47KjQ==.

[7] Álvaro Gómez M. *Diseño de un Sistema de energy harvesting basado en piezoeléctricos.* 2018. URL: https://ebuah.uah.es/dspace/bitstream/handle/10017/33621/TFG_Gomez_Molina_2018.%20pdf?sequence=1.

[8] Rudolf F. Graf. *Modern Dictionary of Electronics.* Newnes, June 1999. URL: https://books.google.com.mx/books?id=o2I1JWPpdusC&pg=PA163&dq=graf+dictionary#v=onepage&q=graf%5C%20dictionary&f=false (visited on 01/15/2023).

[9] *Investigation Of Electromagnetic Emissions In A Deep Underground Mine.* stacks.cdc.gov. URL: https://stacks.cdc.gov/view/cdc/9011 (visited on 01/15/2023).

[10] Zhuo-Wei Miao et al. "A 400-GHz High-Gain Quartz-Based Single Layered Folded Reflectarray Antenna for Terahertz Applications". In: *IEEE Transactions on Terahertz Science and Technology* 9 (Jan. 2019), pp. 78–88. DOI: 10.1109/tthz.2018.2883215. (Visited on 06/30/2019).

[11] *Microbalanza de cristal de cuarzo.* Wikipedia, Jan. 2023. URL: https://es.wikipedia.org/wiki/Microbalanza_de_cristal_de_cuarzo (visited on 01/15/2023).

[12] Antwi Nimo et al. "A new family of passive wireless RF harvesters based on R-C-Quartz oscillators". In: *2013 European Microwave Conference.* 2013, pp. 511–514. DOI: 10.23919/EuMC.2013.6686704.

[13] Técnico Victores. *Esquema del tester para comprobar cristales de cuarzo.* AMYTRONICS HELP, Dec. 2013. URL: https://amytronics.com/como-comprobar-el-cristal-oscilador-en-un-circuito/ (visited on 01/15/2023).

A cluster based on 8 bit, flash based architecture microcontrollers

D L Sánchez-Cabadas[1,2,3], G A Yáñez-Casas*[,1,2], J J Hernández-Gómez[1] and C Couder-Castañeda[1]

[1] Instituto Politécnico Nacional, Centro de Desarrollo Aeroespacial, Belisario Domínguez 22, 06010, Mexico City

[2] Instituto Politécnico Nacional, Unidad Profesional Interdisciplinaria en Ingeniería y Tecnologías Avanzadas, 2580, 07340, Mexico City

[3] ALTEN INGENIERÍA MEXICO SA DE CV, Avenida Antea 1032, Torre Cápita, Piso 11, Interior 76, Jurica

*Corresponding author

E-mail: g.yanezcasas@gmail.com

Abstract. Distributed systems are a suitable alternative to optimize the current embedded technologies, since this kind of systems increase processing power to execute tasks, which otherwise, would take plenty of time to finish with a single processing core, likewise, a larger amount of electronic sensing devices can be added to the system, improving its capacity to collect and analyze data whatsoever. In this work, we propose to use distributed systems theory, in 8 bit CMOS (Complementary Metal-Oxide Semiconductor) microcontrollers, more in particular, devices PIC16F887 from Microchip, which have a reduced and limited architecture. By doing this, a low-power embedded system focused on performing multiple operations and tasks is proposed, in which, time is a vital element, such as real time operating systems, using hard-wired embedded communication protocols to accomplish the correct distribution of data. Using cluster system of the microcontrollers, instead of a microprocessor-based distributed system or a single core embedded system, implies important advantages, such as considerable cost reduction, size flexibility to install this kind of systems in several applications, and a great possibility to modify the initial design and functionality to couple to all the needs of the user. Also, one of the objectives of this development, is to create an easily scalable distributed system, depending on the needs of the use given. The combination of the features mentioned, makes this system suitable for educational purposes, since its simplicity is ideal for students of topics regarding distributed systems.

1. Introduction

The use of distributed systems, brings with it plenty of practical applications in comparison with a single processor computing unit, such as larger storage capabilities, faster processing power, and in most of the cases, the cost of a high-end single core computer to reach certain level of performance is more expensive than a cluster of several low-end computing units.

Normally, distributed systems are applied in the context of computer networks, where each computing element has a microprocessor as a processing unit [1], making them capable of performing general use applications. Its is important to mention that it is not common to relate this kind of systems to embedded systems.

Embedded systems are vastly used in scientific and engineering fields, usually used for small computing applications or specific and repetitive tasks, and currently, this kind of systems are increasing in level of importance as humans keep on automating basic tasks. This embedded systems, have a main processing unit, and some of them, use microprocessors as the core of the whole system, but, they are commonly formed by a microcontroller which has peripherals to read information from sensors, different interfaces for communication with other electronic devices, using either an analog to digital converter or some kind of embedded protocol module, in combination with a software that performs specific functions [2].

To provide some background for this kind of systems, we can take a look at a microcontroller distributed system that was made in [3] in which I^2C protocol [4] was used to communicate with electronic devices with different architectures and, establishing a basis for an heterogeneous distributed system, developing the corresponding protocols conversions to I^2C.

Also, in [5] it is described the process of developing a distributed system for process and condition monitoring. The implemented system uses PIC technology from Microchip [6] alongside an internet connection, resulting in an affordable and flexible structure for monitoring systems.

Another example is a microcontroller-based distributed system that was used in [7] using real-time for the production of cement with a high reliability, more in particular, a group of microcontroller slaves are connected to a central master which is in charge of managing the real time operations of the system.

It is important to notice that in [8] it is described an open architecture microcontroller based distributed measurement and control system with automatic generation of application program, in which a microcontroller cluster was also used to develop the system, using 8bit Atmel microcontrollers of 89S [9] and 89C series [10].

A perfect example of using a distributed system based on microcontrollers to improve processing power on a certain environments is shown in [11]. In this work, the authors exhibit the results of a controlled environment managed by a distributed system of microcontrollers, to control four important parameters for the fish in order to have a higher growth rate.

One of the advantages of embedded systems against microprocessor computing-based units is their power consumption, since microcontrollers are low-power electronic units, and as mentioned in [12] where the authors use a multi-microcontroller distributed system to study power consumption, they mention that microcontroller-based embedded systems are usually powered by batteries, where energy consumption is vital for some applications, which couldn't be done by microprocessor-based computers without consuming a considerable amount of energy.

Also, as mentioned in [13], there are some operations or tasks that are not convenient for master device to do, e.g. real-time control, or input/output operations and which can be done without any problem by a group of slave microcontrollers that perform low-level tasks.

Due to the great potential that microcontrollers represent as processing structures, this paper deals with a cluster design based on microcontrollers that are seen as a flexible and low-cost opportunity for the construction of small-scale distributed systems.

An 8 bit microcontroller system is considered to be implemented, since this kind of architecture generally needs less power demands than the longer word size microcontrollers (16 & 32 bit word size), this happens because in general the clock speed of this microcontrollers is usually faster than the 8 bit microcontrollers, generating a greater power consumption.

Articles regarding distributed systems based on microcontroller cluster arrays are not as common as we anticipated, in those articles, the systems described didn't have a general-use system perspective, in fact, there was only found one written work, in which PIC microcontrollers were exclusively used to form a distributed system [5], and this system was focused on a single task, unlike the objectives of the development shown in this paper, being one of the main goals, to develop a distributed, scalable and general-use system, to increase significantly the processing power of it, if it is required by a certain necessity, depending of the use given to this system.

This work is made up of the following sections: Section 2 describes the electronic and logical elements for the cluster design; in Section 3 the physical circuit is placed and described together with the tests carried out, in Section 4 the conclusions are found and finally in Section 5 the future work.

2. Materials and methods

In this section, the elements used in the development of the microcontroller cluster are described, first, the description of the main microcontroller, the communication protocol and the topology, are placed in order to provide context for the creation of the method used to establish the basis for the distributed system, and then, the method itself to test the system is described, which is the saturation algorithm used.

2.1. Microcontroller PIC16F887

The 8 bit microcontrollers features several embedded functions, such as high-performance and power-efficient processing, a wide range of integrated specific peripherals, from analogical to digital converters, to hard-wired microcontroller protocols, useful for medical, motor control, automotive or even home appliances and applications [14]. Specifically speaking, the Microcontroller PIC16F887 is a 40 pin, 8192 words flash-based, 8 bit microcontroller with a maximum operating frequency of 8 MHz, a 368 bytes Static Random Access Memory (SRAM), an Erasable Programmable Read Only Memory (EEPROM) of 256 bytes, and also a master synchronous serial port for intercommunication applications with other embedded systems [15].

This microcontroller was selected due to its low-end 8 bit architecture and low-power consumption compared to other microcontrollers with a 16 or 32 bit architecture.

2.2. Communication protocol

For the distribution of information, 3 different hard-wired embedded protocols were considered to implement in the microcontroller cluster, the UART [16],SPI [16] and I^2C [4] protocols.

The Universal Asynchronous Receiver Transmitter (UART) module [16] is one of the serial modules available in the PIC device family. The UART is a full-duplex, asynchronous communication channel that communicates with peripheral devices and personal computers [17], using two communication lines (Rx and Tx), allowing communication with only one device per module. The microcontroller PIC16F887 only has one UART module [15], being not able to manage a multi-microcontroller system.

The SPI (Serial Peripheral Interface) module [16] features a synchronous serial data communication bus. The devices in this protocol communicate via a master-slave environment, where the master device starts the communication, and also, the slave devices are controlled by the master through a slave select (SS) line.

As mentioned, the SPI protocol uses a master-slave structure. If the device is configured as the master, it provides and controls the clock signal. All slave devices attached to the master are controlled by this master clock. The master, through the Slave Select (SS) signal, controls the slave device with which it is going to communicate. [18].

This SS increase the number of physical communication lines, one for each slave device, this is the reason why this protocol was discarded from the system.

The Inter-Integrated Circuit (I^2C) [4] module is a serial interface useful for communicating with other electronic or Microcontroller Units (MCUs) (figure 1). The external peripheral devices may be serial erasable programmable read only memories, liquid crystal display drivers, sensors, etc. The I^2C module can operate as three different modes in the I^2C system:

- Slave device.

- Master device in a single master system.
- Master or slave device in a multi-master system.

The I^2C module in microcontrollers contains an independent I^2C master logic and a I^2C slave logic, which generates interrupts based on their events.[19].

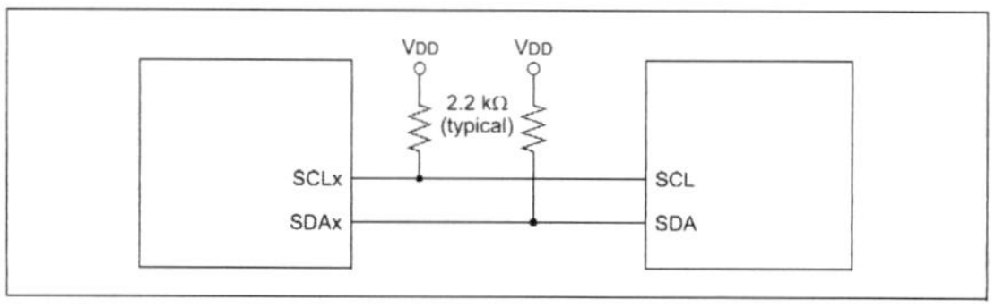

Figure 1. Typical I^2C interconnection block diagram [19]

As observed in the figure 1, the simplest form for the connections in the I^2C protocol is shown, which consists of two signal lines, the SDA (Serial Data) line is in charge of the data transmission, and by the other side, the SCL (Serial Clock) line carries the clock signal, vital for the correct operation of the protocol.

Finally, due to the I^2C protocol capability to manage over 128 devices, and to communicate with all of them on a single communication bus (unlike the SPI protocol) we decided to use this protocol for the correct distribution of the data within the system.

2.3. Topology

There were considered four types of topology to use for the distributed system which are bus, ring, star and mesh[20]. The I^2C protocol limits the system to only use a bus topology or a hybrid that necessarily uses a bus topology, since the protocol communicates with every single device with only one communication bus, formed by a clock line and a data line communicating to each slave device with a unique address programmed by the user or the manufacturer if it is the case.

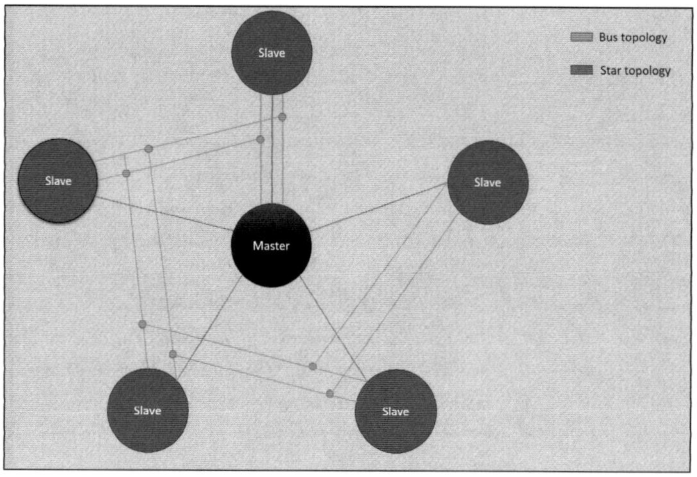

Figure 2. System topology

For the system in question, a bus-star hybrid topology was used to implement the system (figure 2).

2.4. Saturation algorithm

In order to verify cluster capabilities, it is necessary to elaborate an algorithm which saturates the microcontroller or microcontrollers, with a task to be done, to measure the time to finish this task and in order to establish a reference point to compare the performance of a single microcontroller against a microcontroller cluster, this saturation algorithm needs to be implemented first in a single controller unit, then, its time to execute the algorithm it's measured and after that, the same process it's necessary to be done with the entire microcontroller cluster, dividing the whole task, in smaller tasks to be done by the cluster slaves.

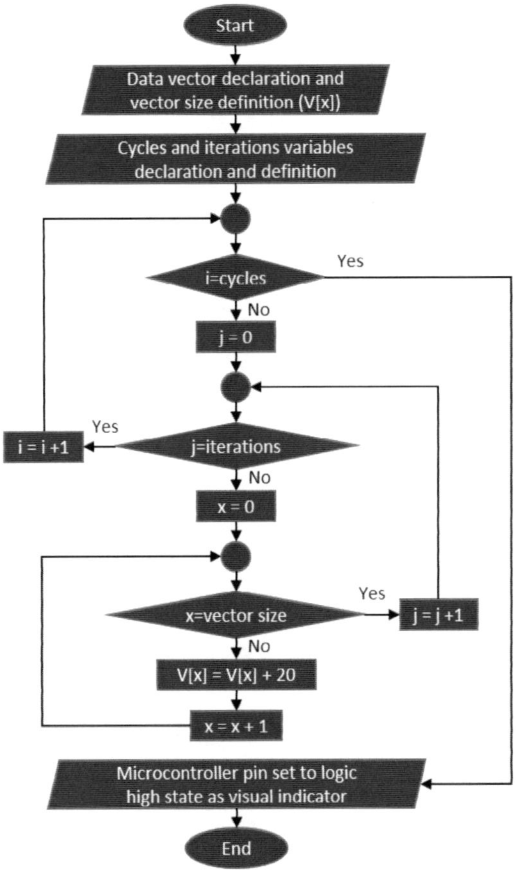

Figure 3. Algorithm flow diagram

This algorithm is written in C language, and before the explanation, it is important to mention the programming limits of the PIC16F887, as it is not possible to declare and process an array bigger than 97 bytes, coming to this conclusion after a process of trial and error, noticing that this happens due to its SRAM architecture and organisation [15].

The algorithm is formed by nested loops. Each one of these, performing a simple task, taking the value of a location of a 95 bytes vector, changing it by adding another value, and storing it on its original location, this process is repeated depending on values of a couple of variables named *iterations* and *cycles*, which control the number of the algorithm's executions. The described process is shown in figure 3.

As soon as the algorithm is solved, a pin changes its state from '0' logic to '1' logic, this is done to have a visual indicator for the PIC to confirm the end of the process.

2.5. Distributed system cluster design

Two simulations of a PIC16F887 microcontroller cluster were made to validate the correct interconnection between all the elements of the circuit, and also, to verify a successful implementation of the I^{2}C protocol [4] with the help of some logic-state pins.

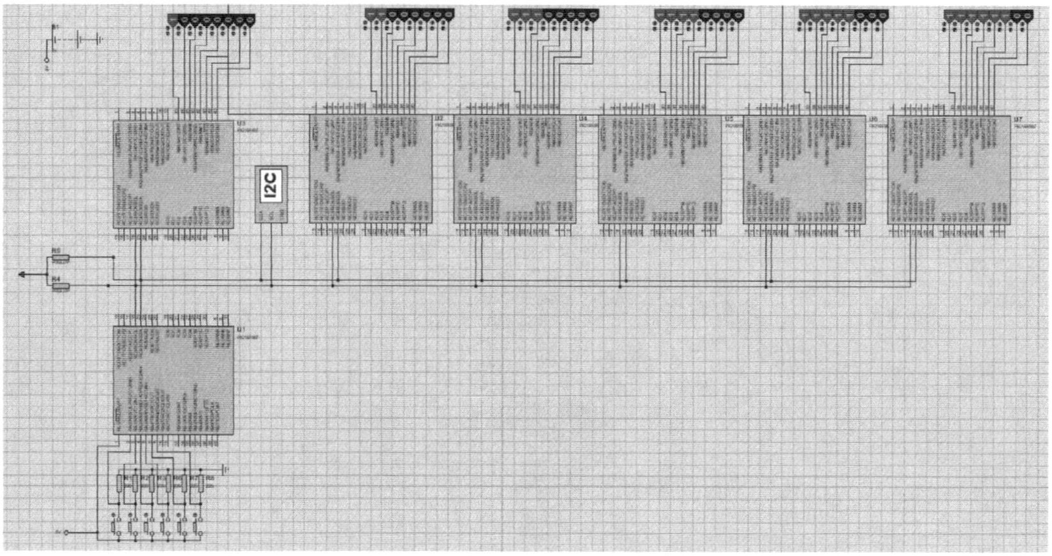

Figure 4. Initial PIC16F887 cluster simulation

First, an algorithm to send data to every single slave from the master and also to receive information from all the slaves to the master was implemented to confirm the adequate distribution of information in the cluster.

Then another simulation was made to measure processing time with a special algorithm, explained in the next subsection, to saturate the microcontrollers using light-emitting diodes and one button.

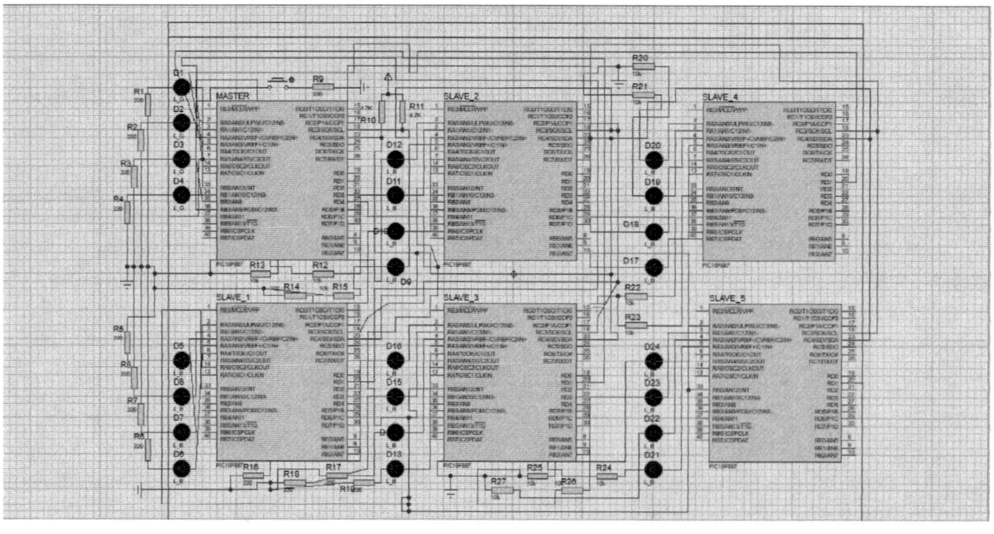

Figure 5. PIC16F887 Cluster simulation to measure processing time

At the beginning in figure 4 is shown the initial PIC16F887 cluster simulation, which contains 1 master device, and 6 slaves devices, each slave device has connected to one of their ports, 8 logic state pins, emulating an 8 bit binary number. This number is send to the master device by means of a button from each one microcontroller slave, and after the reception of the number, the microcontroller immediately shows the corresponding value in one of its ports, also, an I^2C debugger is connected to the I^2C bus to verify the correct data transmission.

Then, in figure 5, is the corresponding simulation to the processing time test, which consists in one PIC16F887 master microcontroller and five PIC16F887 slave devices, but this time, each slave has 4 light-emitting diodes (LED) in one of its ports, which show the state of the data distribution process. which goes as follows, first, the user pushes the button connected to the master device, then, the master distributes an 95 bytes long vector, sending 19 bytes to each slave, and at this point, one LED is turned on in each slave, to indicate the correct reception. Second, once the data bytes were sent to all the slaves, each one of them (slaves), start executing the saturation algorithm, and once they have finished, another two LED's are turned on, and a signal is sent to the master, indicating that the algorithm was successfully solved. Finally, until every single slave has sent the confirmation signal, the master requests the processed data from each slave, gathers all the data bytes, and turns on a whole port, indicating that all the process has finished.

3. Results and Discussion

Once the simulations were checked and validated, a physical circuit was built to verify both the interconnection between the microcontrollers with the I^2C protocol and, also, the saturation algorithm operation.

As the saturation algorithm is a controlled process, it is only until the microcontroller finishes processing that loop, that it turns on an indicator, so that the user knows when the algorithm is solved, and when the microcontroller is not saturated whit instructions to execute.

First, after the communications protocol was verified physically, the saturation code was implemented in a single microcontroller (figure 6). The circuit was powered with 5V using a voltage source, and every electronic element was placed on 3 coupled proto-boards, this circuit is the implementation of the simulated circuit made in figure 5, thus, all of the mentioned electronic elements are the same as the ones used in the simulation.

The processing time was measured using a digital timer, being 64.5 seconds for the microcontroller to solve the algorithm.

Then the distributed system was tested, implementing a code which has both, the saturation algorithm and the I^2C communication algorithm [4] to distribute the information to process, to each slave in the cluster.

Initially, the test was made with 6 microcontrollers, but this system is scalable up to 127 devices which is the maximum amount of devices supported by the I^2C protocol [4], and with this 6 devices, it was possible to make a configuration of 5 processing slave units and 1 master data compiler which distributes the 95 bytes length vector (defined in the materials and methods section) equally, with a 19 bytes buffer size to each slave (95 bytes divided by the 5 slave devices).

The master device was programmed to work at the fastest clock frequency (8 MHz) [15] and also, to operate at the maximum speed allowed for the I^2C protocol for this device, using the equation 1 to determinate 1 MHz as the protocol frequency [15].

$$clock = \frac{FOSC}{(4 \times (SSPADD + 1))} \tag{1}$$

Being "FOSC" the oscillation frequency of the microcontroller clock (FOSC = 8000000), and "SSPADD" a value determined by the user, which in this case is SSPADD = 1.

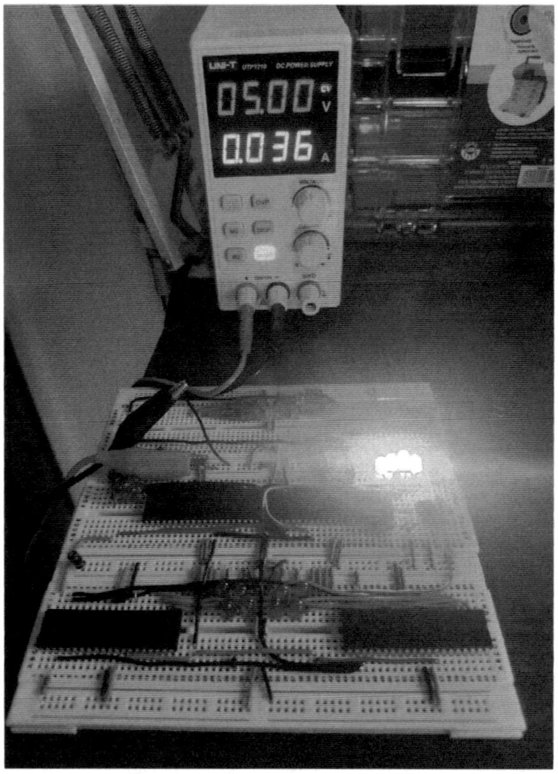

Figure 6. Physical cluster circuit with a single microcontroller solving the saturation algorithm

Figure 7. Physical cluster circuit with 6 microcontrollers solving the saturation algorithm

In the figure 7 is the final implementation of the distributed system, also powered with a 5 V voltage source, in which every microcontroller has LED's (green LED's for the master device and blue LED's for the slave devices) as indicators to show the state of the system, and as mentioned in figure 5 the same process mentioned in the simulation processing time test is executed, and when finished, all the indicators are turned on, letting the user know that the task is done.

The time of the whole process was measured using a digital timer, including processing time and transmission time, obtaining 27.1 seconds to solve the algorithm and to gather all the data back to the master microcontroller.

Finally, a Printed Circuit Board (PCB) was designed, shown in figure 8, using the Plastic Dual In-line Package (PDIP) for the PIC16F887 devices [15], Surface Mounting Device (SMD) format for all the light-emitting diodes. To the left is the two dimension design, and to the right, the 3D virtual PCB model.

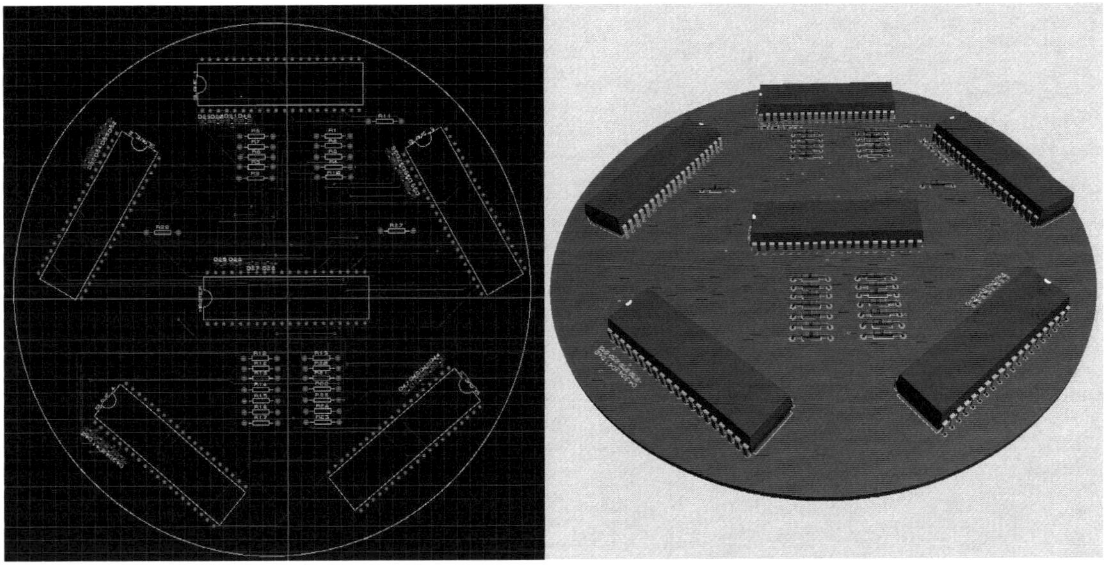

Figure 8. Printed circuit board design

When the system is executing the main saturation algorithm, the total current of the whole system measured is 0.30 A, but in order to obtain the real current consumption of the system, regardless of the visualisation interface, the current demanded by the LED´s must be subtracted in the calculation, each LED requires 0.01515 A due to the fact that the system is powered with 5V and that each LED is connected to a 330 ohm resistor (5V/330 ohm), and since there are 7 LED´s turned on at this stage, then the current demanded by the LED´s is 0.10606 A (0.01515 A x 7), finally, doing the corresponding subtraction (0.30 A - 0.10606 A), the current consumption of the system while executing the code is 0.1939 A. The current consumption of the system measured when it is at an idle state is 0.0035 A.

4. Conclusions

The microcontroller-based distributed system obtained lays a foundation for future development focused on performing embedded tasks more efficiently.

Due to its great design flexibility, it can be used to reduce processing time for a complex algorithm, also, it can be used to optimize embedded real time systems, since this microcontroller cluster has more controller units, it is able to read less sensors for each microcontroller, unlike a single unit reading several sensors at the same time, leaving enough time to process and send the information to a central unit, improving sensor processing time.

This distributed system is capable of managing up to 127 slave devices, increasing the processing power of the system for each unit.

Considering that a single microcontroller can process 95 bytes in 64.5 seconds, and since every MCU is programmed to operate at the same clock frequency, this means that each microcontroller is capable of processing 1 byte in 0.678947 seconds (95 bytes/64.5 seconds).

Since the microcontroller cluster executes both the saturation, and the data transfer algorithm in 27.1 seconds, and inasmuch each microcontroller processes 19 bytes for each of the 5 slave microcontrollers, then, the processing time and the data transmission time can be inferred, being 12.9 seconds to process the data bytes (0.678947 seconds per byte). Due to the fact that there are 5 microcontrollers, the system processes 5 bytes in that time (0.678947 seconds), which consequently causes 95 bytes processed in 12.9 seconds, from that consideration, the data transmission time is 14.2 seconds (27.1 - 12.9 seconds) and considering that the data transmission time in any case is 14.2 seconds, then, the equation 2 is used to determine the approximate time to finish processing the whole algorithm, depending on the amount of microcontrollers interconnected in the system.

The results are shown in the table 1.

$$T = \frac{95 \times 0.67894}{(MCU's)} + 14.2 \tag{2}$$

Being T the time to execute the algorithm and $MCU's$ the number of microcontrollers in the cluster.

Table 1. Algorithm time execution estimation

Number of microcontrollers	Time to execute the algorithm	Processing effectiveness against a single microcontroller configuration	Time saving against a single microcontroller configuration
1	64.50 Seconds	-	-
6	27.10 Seconds	57.98%	37.40 Seconds
10	20.65 Seconds	67.98%	43.85 Seconds
20	17.42 Seconds	72.99%	47.08 Seconds
50	15.49 Seconds	75.98%	49.01 Seconds
95 - 127	14.87 Seconds	76.94%	49.63 Seconds

It is important to note that from 95 to 127 microcontrollers the time is the same, this happens because when there are minimum 95 microcontrollers in the system, the 95 bytes array can't be divided any more, since the bytes necessarily have to be integers.

Regardless of the kind of the saturation algorithm, and due to the fact that in this 8 bit microcontroller, the maximum number of bytes on a single array is 95, as a result, this array can only be separated in 95 bytes maximum, thus, 95 microcontrollers are the limit for this system.

For the calculation of the processing effectiveness, 64.5 seconds were considered to be 0%, and 0 seconds were considered to be 100% of effectiveness, being 76.94% the maximum processing effectiveness reachable by this distributed system compared to a single microcontroller processing the algorithm.

As a consideration, another algorithm to measure the processing effectiveness of the microcontroller cluster could be written, only this time, it could be based on executing multiple and more complex operations, using double type variables.

below, in the table 2 is deployed the cost for the construction of the physical structure of the prototype.

Table 2. Electronics cost

Component name	units	Price per unit (USD)	subtotal
PIC16F887	6	4.63	27.78
Push button	1	4.91	4.91
LED	24	0.033	0.792
Protoboard	3	7.49	22.47
Connection cable	1	10.52	10.52
Electrolytic capacitor	1	0.71	0.71
		Total	61.182

As observed on this table, this is an economical and affordable distributed system, this opens the possibility for further development of cheaper, and more effective embedded systems for more complex projects, involving data obtaining from sensors, sampling signals, or processing information stored on a digital memory.

Also, it is important to mention that this system is flexible and versatile, since this architecture allows the electronics circuit to be easily modifiable, as well as the communications protocol, also, the interface to visualise the state of the system could be changed without any relative difficulty. This exact modifications would be more difficult to implement in a longer word size microcontroller since there are more registers and parameters to modify in order to achieve this changes.

From a didactic point of view, this system could be used for teaching reasons, helping early-stage distributed system students to implement scalable distributed systems.

5. Future work

The microcontroller-based distributed system obtained has a lot of potential to be used and deployed in different engineering branches, such as aerospace technologies, used as an alternative of a on-board computer, so, a distributed system with a larger amount of microcontrollers could be designed and implemented, also, a sensor monitoring algorithm can be designed to be integrated to the system, in order to reduce processing power to the master microcontroller, giving it more time to execute safety checks or managing important modules of certain sub-systems.

6. Acknowledgements

The authors acknowledge partial economical support by ALTEN INGENIERÍA MEXICO SA DE CV and by projects 20220907, 20222032, 20221182 and 20220176, as well as EDI grant, provided by Secretaría de Investigación y Posgrado, Instituto Politécnico Nacional, also

References

[1] F L 2015 *Sistemas Distribuidos* (UAM colección una decada) ISBN 9786072804760 URL `http://dccd.cua.uam.mx/libros/texto/texto_pdf.php?url=LopezFuentes-SistemasDistribuidos`

[2] White E 2011 *Making Embedded Systems: Design Patterns for Great Software* Oreilly and Associate Series (O'Reilly Media, Incorporated) ISBN 9781449302146 URL https://books.google.com.mx/books?id=bI8w17SyNdYC

[3] Sastry J, Ganesh J V and Bhanu J S 2015 *Indian Journal of Science and Technology* **8** cited by: 19; All Open Access, Bronze Open Access

[4] Phillips Semiconductors 2003 *AN10216-01 I2C Bus* Phillips Semiconductors Eindhoven, Netherlands an10216 ed

[5] Frankowiak M R, Grosvenor R I and Prickett P W 2005 *Microprocessors and Microsystems* **29** 189 – 196

[6] Ibrahim D 2014 *PIC Microcontroller Projects in C: Basic to Advanced* (Elsevier Science) ISBN 9780080999678 URL https://books.google.com.mx/books?id=xQajAgAAQBAJ%7D

[7] Cameron R and Zhao J 1993 *Control Engineering Practice* **1** 441 – 447

[8] Bolic M, Drndarevic V and Samardzic B 2001 *Sensors and Actuators, A: Physical* **90** 215 – 221

[9] Atmel Corporation 1997 8-bit Microcontroller with 12 Kbyte Flash AT89S8253 https://ww1.microchip.com/downloads/en/DeviceDoc/doc3286.pdf Online. Last accessed Sep 2022

[10] Atmel Corporation 1997 AT89C2051 https://ww1.microchip.com/downloads/en/DeviceDoc/doc0368.pdf Online. Last accessed Sep 2022

[11] Neelamegam P, Kumaravel S and Raghunathan R 2008 *Instrumentation Science and Technology* **36** 515 – 524 cited by: 2 URL https://www.scopus.com/inward/record.uri?eid=2-s2.0-48849107153&doi=10.1080%2f10739140802234964&partnerID=40&md5=9f8b4f5cbdc8ec293905c381820f08f3

[12] Szymczyk M and Augustyniak P 2022 *Electronics (Switzerland)* **11** cited by: 0; All Open Access, Gold Open Access

[13] Netland nd Skavhaug A 2016 *Lecture Notes in Computer Science (including subseries Lecture Notes in Artificial Intelligence and Lecture Notes in Bioinformatics)* **9923 LNCS** 256 – 267

[14] Microchip Technology Inc® 2022 8-bit PIC and AVR Microcontrollers https://ww1.microchip.com/downloads/aemDocuments/documents/OTH/ProductDocuments/Brochures/8-bit-MCU-Brochure-30009630.pdf Online. Last accessed Sep 2022

[15] Microchip Technology Inc® 2015 *PIC16F882/883/884/886/887 Data Sheet* Microchip Technology Inc.®

[16] Dawoud D and Dawoud P 2020 *Serial Communication Protocols and Standards: RS232/485, UART/USART, SPI, USB, INSTEON, Wi-Fi and WiMAX* River Publishers Series in Communications Series (River Publishers) ISBN 9788770221542 URL https://books.google.com.mx/books?id=nj50zQEACAAJ

[17] Microchip Technology Inc® 2007 PIC24F Family Ref. Manual, Sect. 21 UART https://ww1.microchip.com/downloads/en/DeviceDoc/en026583.pdf Online. Last accessed Sep 2022

[18] Microchip Technology Inc® 2018 Serial Peripheral Interface (SPI) Communications on 8-Bit PIC® Microcontroller https://ww1.microchip.com/downloads/en/Appnotes/TB3157%20Serial-Peripheral-Interface-Communications-on-8-bit-PIC-90003157A.pdf Online. Last accessed Sep 2022

[19] Microchip Technology Inc® 2015 dsPIC33/PIC24 FRM, Inter-Integrated Circuit (I2C) https://ww1.microchip.com/downloads/en/DeviceDoc/70000195g.pdf Online. Last accessed Sep 2022

[20] Tanenbaum A S 2003 *Computer Networks* (*Computer Networks* no p. 3) (Prentice Hall PTR) ISBN 9780130661029 URL https://books.google.com.mx/books?id=dRNLAQAAIAAJ

Proposal of rectennas as an auxiliary device for the electrical power subsystem of a 1U CubeSat

I Medina Sánchez[1], **A Flores**[1] **and E E Lozano**[1]

[1]Instituto Politécnico Nacional, Escuela Superior de Ingeniería Mecánica y Eléctrica, unidad Zacatenco, Mexico City, Mexico

E-mail: ismesa@gmail.com

Abstract. Current electrical power subsystem (EPS) devices operate solely on solar panels and batteries, depending on them to provide at least 5 watts of power for a 1U CubeSat, this research proposes an auxiliary method to increase the CubeSat power or lessen the battery sizes by providing a different energy source via a rectenna. These devices are capable of delivering an output voltage by harnessing electromagnetic waves in its surroundings, there are different types of circuits and this paper compiles the best candidates to operate in a 1U CubeSat. Nanosatellites are strongly restricted by their weight and energy usage, thus by using an auxiliary device for the EPS that could potentially reduce the size of said subsystem, the results are an even more efficient CubeSat in terms of energy and space distribution.

1. Introduction

The electrical power subsystem of a satellite (EPS), and specifically a CubeSat, is formed by solar cells and batteries that supply electrical energy to the circuits; during the orbit of a satellite, it enters an eclipse state, where there is no direct solar light incidence, this is when the rechargeable batteries supply the energy to the system, it is worth mentioning that there are also electrical regulator devices that help to prevent pulses introduced into the power lines towards the payload, thus avoiding the deterioration of the cells [6].

In artificial satellites, solar cells that are used as an electrical power source and are constantly exposed to the bombardment of charged particles of various energies. This causes defects in the materials of the cells, deteriorating them and, therefore, can cause important losses in their electronic properties [12].

Regarding batteries, there are different reasons why they could usually fail, some are due to poor design or manufacturing defects, but it is true that there are also external factors, such as overheating, recurrent shocks, perforations or problems with the charger, which can cause a short circuit or increase the temperature of the cell until the internal separator of the battery breaks [11].

In this way, knowing some of the issues that arise in the existing EPS in a CubeSat, the proposal consists on the implementation of rectennas as an auxiliary EPS method for a CubeSat, which would be used as a backup to both the panels and the batteries. It is imperative to have a backup system and not relying entirely on the solar panels that could not generate an optimal electrical signal all the time for different stages of the orbit, or the batteries that have a very limited usage and could present damage to the EPS. For these cases, the CubeSat could benefit by using a type of antenna that is characterized by the direct conversion of microwaves into direct current, i.e. rectennas, that would support the energization of a CubeSat [1].

2. Development

This section shows the figures of the circuit proposals for the rectennas, followed by the simulation results developed in Proteus. It is important to remark that as simple as these circuits are, the tunning of the frequency is an unresolved issue, this is the reason why only simulations were carried out for this work that is currently in progress.

2.1. Circuit 1 (CT1)

Figure 1 shows a circuit that consists of a rectenna that coexists with quartz crystal, when receiving a signal, the crystal is excited and due to its piezoelectric properties, it generates a voltage and a higher frequency wave. When passing through the Schottky diodes, the complete wave is rectified and thus it generates a direct current voltage which is sufficient to feed a CubeSat, however, one of the limitations of this circuit is that the quartz must oscillate at the exact frequency in order to obtain the necessary output voltage.

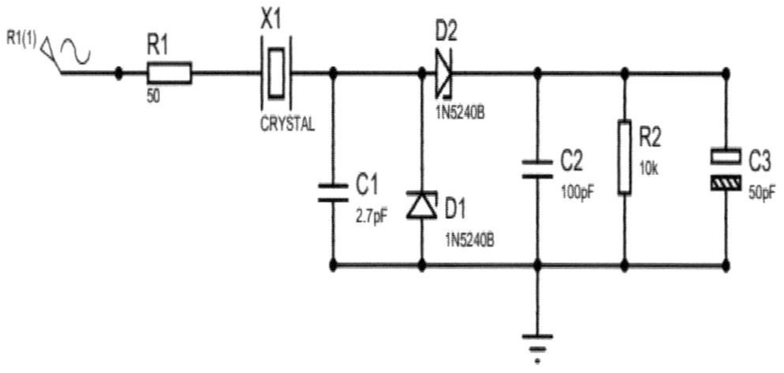

Figure 1. Rectenna with quartz oscillator [8].

2.2. Circuit 2 (CT2)

Figure 2 presents a very easy to construct circuit, the most basic rectenna, where basically the received signal only passes through a capacitor and a resistor that are in parallel which is intended to rectify the wave, however, although the voltage output would be sufficient to feed a CubeSat, these are highly unstable devices and do not usually achieve or complete their mission and therefore it is not so reliable to implement it in a CubeSat.

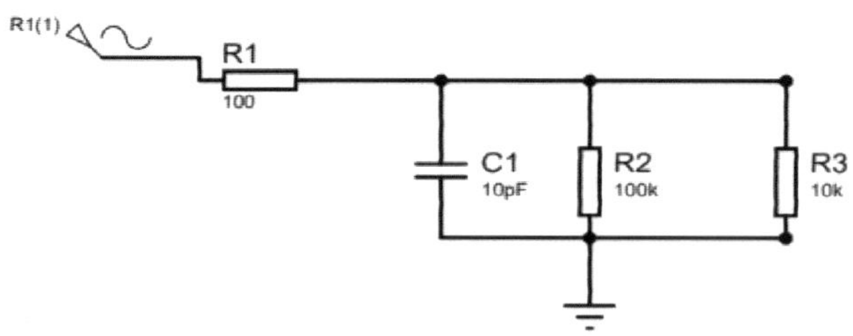

Figure 2. Simple rectenna [3].

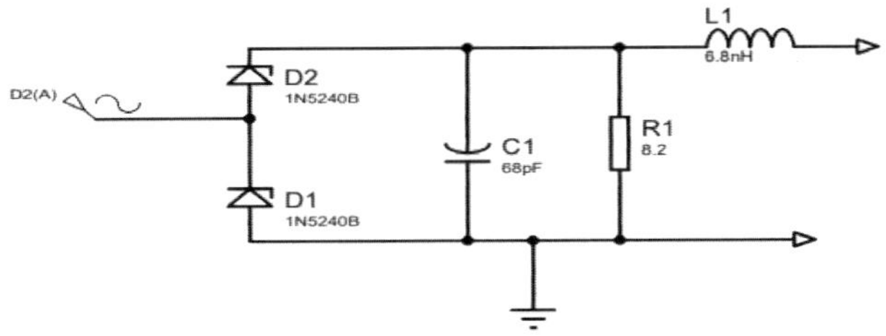

Figure 3. 3 Rectenna with coil [5]

2.3. Circuit 3 (CT3)
Figure 3 shows a circuit which goes through a rectification stage of its signal at the input and coexists with a coil. It is possible to obtain sufficient voltage to satisfy the EPS as well as obtaining a complete wave rectification, however, to achieve the correct operation, the coil must be built according to the necessary frequency which makes it difficult to elaborate, because the fine tuning is very precise and, if not done properly, it will not generate the necessary voltage for the CubeSat.

2.4. Circuit 4 (CT4)
Figure number 4 shows a circuit that passes through a diode bridge which is useful for rectification of the full frequency wave, where the capacitors help to have some energy storage in the circuit and, at the time of presenting voltage losses, these serve as support; however, the bandwidth in which the circuit works is low since, at the time of setting a high frequency, the voltage output decreases and with relatively low frequencies the output voltage is more than enough to power a CubeSat.

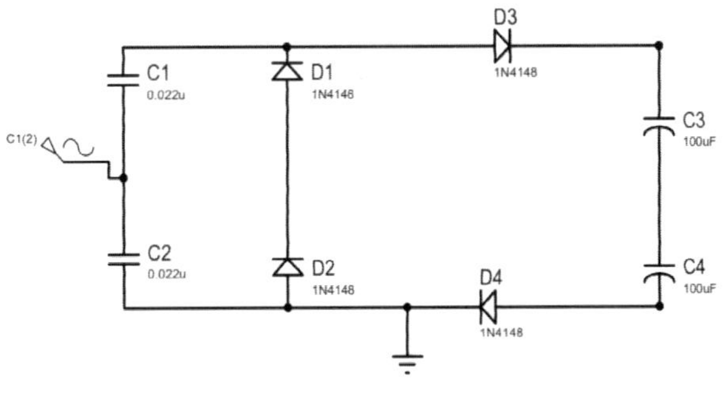

Figure 4. Rectenna diode bridge [2] .

2.5. Circuit 5 (CT5)
Figure 5 shows the circuit in which the wave is completely rectified. When the input voltage is negative, the current passes through D2 and when it is positive it goes through D1, turning off D2, and all its current is accumulated in C2, obtaining the sum or supporting to make the sum of both currents; however, due to its complex operation, a sufficient output voltage is not obtained for the CubeSat but it does have a wide bandwidth.

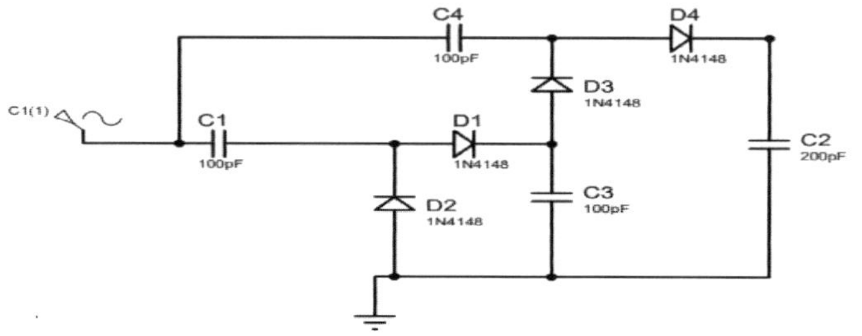

Figure 5. Rectenna with diode array [10] .

2.6. Circuit 6 (CT6)

Figures 6 a) and b) show a circuit that receives an input signal that passes through a transformer that increases the voltage and this is rectified by the diodes, however, the circuit does not provide enough voltage for the CubeSat's EPS with or without the capacitor. Although, with the capacitor, this allows storing a little voltage making the output voltage a little higher. In the same way, this circuit can operate with a wide bandwidth which is very helpful in the space field, however, by implementing the transformer, this is greater robustness compared to other circuits.

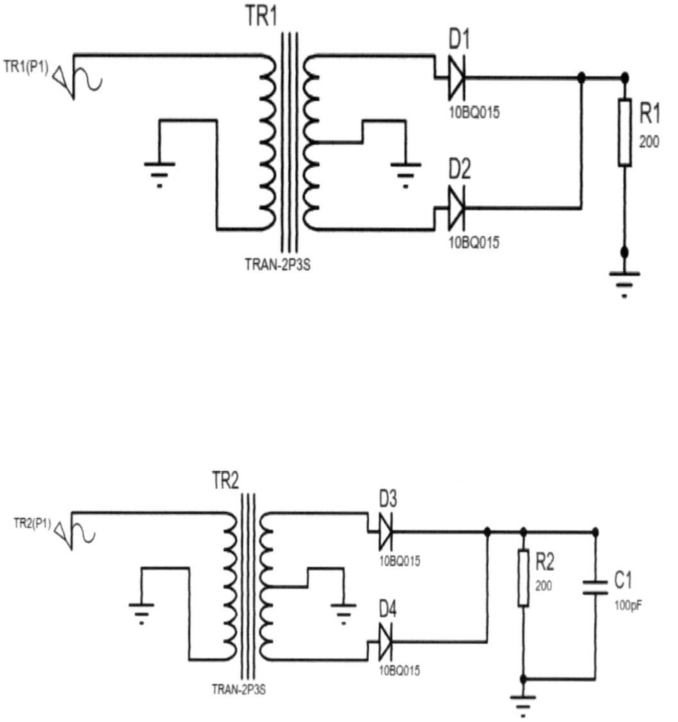

Figure 6. a) Rectenna with transformer. b) Rectenna with transformer and capacitor [7].

2.7. Circuits results and comparison

Table 1 depicts the results of each circuit considering a constant input signal amplitude of 20 V.

Table 1. Input and output values in the circuits.

Circuit	Input amplitude	Output voltage
1	20 V	9.25 V
2	20 V	14.5 V
3	20 V	10.6 V
4	20 V	36.3 V
5	20 V	1.53 V
6a	20 V	1.92 V
6b	20 V	2.75 V

In table 2 a comparison of the previous circuits is made in order to make the selection of the optimal circuits that could be considered for the functionality that is intended to operate as an auxiliary method in a CubeSat's EPS.

Table 2. Input and output values in the circuits.

Model	Bandwidth (400MHz – 3GHz)	Full wave rectification	Easy development	Easily acquired elements	Sufficient output voltage (greater than 3.6 V)
CT1	×	✓	×	×	✓
CT2	×	×	✓	✓	✓
CT3	×	✓	×	×	✓
CT4	×	✓	×	✓	✓
CT5	✓	✓	×	×	×
CT6a	✓	✓	×	✓	×
CT7b	✓	✓	×	✓	×

×- Does not comply
✓-Meets

For the selection of the optimal circuits it has to be taken into account that the output voltage is sufficient, as well as the rectification of its wave is complete, therefore, and based on these parameters, the best circuits proposed for the implementation in the EPS of the CubeSat are the circuits 1, 3, 4. However, as shown in table 2, these do not have a wide bandwidth and could be troublesome to implement due to the fine tuning of the components.

In a CubeSat, different types of frequency are received, and this depends on the mission, orbital position and type of antennas. These frequencies vary from 400MHz to 3 GHz [9].

Generally, the CubeSat is powered with a voltage ranging from 5V to 12V. An example of the batteries used are the Molicel INR18650-P26A which provides 3.6V and 2.47A, this results in 8.9Wh, so it would be needed a battery bank of minimum 4 of them, considering that the size of each of them is 64.8 mm long × 18.4 mm in diameter and a weight of 46g [4], the battery bank would measure approximately 64.8 mm long by 73.6 wide with a weight of 184 g.

A rectenna circuit would only have a maximum measurement of 25 mm long by 25 mm wide, each one with very low weight.

Circuit 1 has a weight of 19.67g which would be 89.3% lighter compared to the battery bank, circuit 3 has a weight of 25.34g, 86.2% lighter than the battery bank and circuit 4 weights 29.21 g, being 84.12% lighter than the battery bank, this results in a great saving in both weight and size of the battery bank, if it was replaced by one of the circuits or reduced in size and obtain the remaining energy via the rectenna.

3. Conclusions

To determine if a rectenna is viable to use in a CubeSat some parameters must be taken into account, such as bandwidth, wave rectification, as well as output voltage, and these must be considered together since it would be useless to comply with the bandwidth if it does not deliver enough output voltage to power the nanosatellite.

The designs of the circuits seem very simple, so their physical construction is uncomplicated, and they can be used frequently and could be considered to use in the design of the CubeSat's EPS. Throughout this research, it has been proved that the use of rectenna as an auxiliary system for energizing a CubeSat would be of great benefit to mitigate the problems that may arise with the current energy subsystem devices of the CubeSat. With the correct use of the rectenna, it could be achieved that the energy subsystem of a CubeSat is not interrupted at all and this in turn would allow the mission for which the CubeSat was made to have greater success.

One of the advantages of including the rectennae to the nanosatellite is the saving of space inside the cubesat as well as the reduction of its weight, which could allow us to take advantage of that space and weight to add some other elements to the CubeSat which make it have greater functionality. Finally, it is worth mentioning that there is a lack of development to have a rectenna that meets all the requirements since for now it can only be observed in simulations, however, this research is a borderline for satellite innovation. Nor could it only be applied in the spatial field, if good development is achieved and good results are presented, it could be implemented in other applications with which they interact in everyday use.

Acknowledgments This project has been partially supported by project 20221182, as well as EDI grant, all provided by Secretaría de Investigación y Posgrado, Instituto Politécnico Nacional.

References

[1] Bryan Javier Blanco-Ibalñez David Santiago - Cerquera-Zambrano. *Diseño y simulación de una rectenna para cosechar energía electromagnética a 2.4 GHz.* Tech. rep. Fundación Universitaria de Ciencias de la Salud, 2020. URL: `https://hdl.handle.net/10983/24502`.

[2] *circuit analysis - Rectenna - What's the Purpose of the Capacitors?* Electrical Engineering Stack Exchange. URL: `https://electronics.stackexchange.com/questions/464069/rectenna-whats-the-purpose-of-the-capacitors`. (visited on 01/15/2023).

[3] Evgeniy Donchev et al. "The rectenna device: From theory to practice (a review)". In: *MRS Energy Sustainability* 1 (2014). DOI: `10.1557/mre.2014.6`. URL: `https://pdfs.semanticscholar.org/1e65/e2fad3d7c03d83131573176e99d542d57758.pdf` (visited on 11/07/2019).

[4] Aldo Eleazar et al. *Baterías que dan vida a un nanosatélite.* URL: https://www.amc.edu.mx/revistaciencia/images/revista/71_3/PDF/07_71_3_1215_BateriasNanosatelite-L.pdf.

[5] A. Gautam and Binod Kanaujia. "A novel dual-band asymmetric slit with defected ground structure microstrip antenna for Circular Polarization operation". In: *Microwave and Optical Technology Letters* 55 (June 2013), pp. 1198–1201. DOI: 10.1002/mop.27547.

[6] Javier Hernández Sánchez and Florentino Jiménez Muñoz. *Sistemas de comunicación por satélite: utilización en los sistemas de navegación aeronáuticos.* oa.upm.es, Nov. 2015. URL: https://oa.upm.es/40769/ (visited on 01/15/2023).

[7] José Mezquita Fayos. *Diseño de antenas compactas para aplicaciones de Energy Harvesting.* riunet.upv.es, May 2015. URL: https://riunet.upv.es/handle/10251/50917?show=full (visited on 01/15/2023).

[8] Antwi Nimo. "A new family of passive wireless RF harvesters based on R-C-Quartz oscillators". In: *www.academia.edu* (Jan. 2013). URL: https://www.academia.edu/67406440/A_new_family_of_passive_wireless_RF_harvesters_based_on_R_C_Quartz_oscillators (visited on 01/15/2023).

[9] Catalina Núñez Lobete. "Diseño de antena en banda UHF para Cubesat". In: *digibuo.uniovi.es* (2018). URL: https://digibuo.uniovi.es/dspace/handle/10651/49025 (visited on 01/15/2023).

[10] Ugur Olgun, Chi-Chih Chen, and John L. Volakis. "Wireless power harvesting with planar rectennas for 2.45 GHz RFIDs". In: *2010 URSI International Symposium on Electromagnetic Theory.* 2010, pp. 329–331. DOI: 10.1109/URSI-EMTS.2010.5637008.

[11] ALDO ELEAZAR PÉREZ RAMOS, E. M. A., and S. V. R. *BATERIAS QUE DAN VIDA A UN NANOSATELITE.* 2020.

[12] *Small Satellites and Their Regulation — PDF — Space Debris — Satellite.* Scribd. URL: https://es.scribd.com/document/350129108/Small-Satellites-and-Their-Regulation (visited on 01/15/2023).

Remote sensing application using Landsat 8 images for water quality assessments

E E Cruz-Montes[1], M M Durango-Banquett[1], F M Torres-Bejarano[2], G A Campo-Daza[2] and C Padilla-Mendoza

[1]Environmental Engineering Program, Universidad de Córdoba, Cra. 6 No. 77-305, Montería, Colombia

[2]Departament of Environmental Engineering, Universidad de Córdoba, Cra. 6 No. 77-305, Montería, Colombia

E-mail: franklintorres@correo.unicordoba.edu.co

Abstract. This study aims to implement remote sensing to determine water quality parameters. Based on the water quality parameters measurements on February 5, 2021, and the LANDSAT 8 satellite images reflectance values, statistical models were generated by the RLM method: Stepwise Regression in Matlab software and digital models through QGIS and ArcGIS Pro GIS. The models obtained for pH, temperature, and turbidity were above 0.6 R2, while dissolved oxygen was above 0.8, showing a good correlation between in situ and estimated data. It is important to avoid cloudy conditions covering the study area because they limit the mathematical models' application, altering or preventing the results from being generated. The implementation results of this type of technology are of great relevance for future projects, where water quality assessments can be made by using national and international regulations according to the use either for human consumption or for aquatic life protection.

1. Introduction

To solve problems of access, economy, and availability of information, remote sensing technologies have been used worldwide, since they allow information collection in a specific place and time for a phenomenon study or object without the need to be in direct contact with it[12] [1] Likewise, water quality assessment through remote sensing mapping of parameters has been implemented for several decades in the United States[8]; also, in investigations to estimate chlorophyll (a) concentration and distribution and total suspended matter respectively for cases in inland water bodies in Jordan or at the Vaal Dam in South Africa northeastern region used Sentinel 2 and LANDSAT 8 satellites to assess water quality [10] [19], also in Bahía Colorada, Mexico for the determination of the variation of the temporal space of phosphates (PO4), electrical conductivity (CE), total suspended solids (TSS), turbidity and pH [7].

Given this advantage, methodologies that integrate remote sensing using satellite images' spectral bands have been integrated for water quality monitoring and multiple linear regression to associate reflectance data with environmental parameters measurements in surface water bodies allowing these parameters' spatial estimation without the need to access the interest area again [24]. The potential use of 30m resolution satellite images (LANDSAT) is available to the public since 2000 [3], to overcome the conventional tracking methodology shortcomings based on the water's optical properties by integrating reflectance, GPS, and GIS data [28], taking into account that water quality parameters are directly and indirectly related to the water's reflectance characteristics, or so-called "optically active surface components" [29], solar radiation reflected at different wavelengths by water bodies can therefore be measured using remote sensors [2].

In Colombia, the regulation on the desirable ranges or maximum permissible limits of water quality parameters for surface water bodies according to use is limited. The Technical Regulations for the Potable Water and Basic Sanitation Sector (RAS) on resolution 1096 of 2000 [16], establish acceptable ranges or limits, dissolved oxygen between them ($> 4ppm$), for water sources surface or underground as a basis for the application of minimum treatment processes for the purification of water for human consumption; likewise, resolution 2115 of 2007 by the Ministry of Environment, Housing and Territorial Development (MAVDT) establishes the physical characteristics for human consumption such as conductivity ($< 1000\mu$S/cm), pH (6.5 -9.0) and turbidity ($< 2FNU$) [17].

Two main problems stand out: pollution processes due to anthropic activities affecting water quality in the Guartinaja, Momil, and Sapal wetlands as part of the Ciénaga Grande del Bajo Sinú and, on the other hand, infrequent monitoring in water bodies affects information availability for water quality assessment. Thus, this research proposes remote sensing use with LANDSAT 8 satellite images to evaluate the wetlands' water quality in the dry season by estimating statistical models by multiple linear regression (RLM).

2. Materials and Methods

2.1. Study area

Bajo Sinú's Ciénaga Grande is located between the Bajo and Medio Sinú subregions of the department of Córdoba – Colombia and is composed of 17 wetlands, including the Guartinaja ($9°, 1314.44"Nand75°, 4513.04"W$), Momil ($9°, 1334.34.34.464"Nand75°, 4045.14"W$) and Sapal ($9°, 1244.42"Nand75°, 3956.55"W$) wetlands as the interest area as shown in Figure 1. The study area is characterized by floodplain-type geomorphology that occupies about 44.000 hectares; it also has a 27.6°C mean annual temperature and 1.350 mm multiannual precipitation [5].

La Ciénaga is formed by the dynamics of the Sinú River in the lower area of the basin, between the municipalities of Chimá, Cotorra, Lorica, Momil, and Purísima and represents the basis of the socioeconomic, cultural, ecological, and environmental prosperity of neighboring sectors [22] [27]. Some of the most important activities that take place in the swamps are fishing, agriculture, and livestock; In the same way, the construction of irrigation districts, embankments, and dikes has been carried out, coupled with a population increase, pressures are generated not only on the water resource but also on the ecosystem dynamics between species [26].

2.2. LANDSAT 8 description

Landsat-8 satellite was launched on February 11, 2013, by the National Aeronautics and Space Administration (NASA) and the United States Geological Survey (USGS) [21]. The LANDSAT 8 satellite can measure the electromagnetic spectrum frequency ranges through two main sensors, considering that each range corresponds to a band, i.e., a color that may not be visible [14].

2.2.1. Sensor Operational Land Imager (OLI). The images are characterized by 12 bits radiometric resolution and a 30 meters spatial resolution for the visible (RGB) and near-infrared and shortwave (NIR) spectrum bands except for panchromatic band 8, which has 15 meters of spatial resolution; these images are captured over a 185-kilometer-wide area [13]. Currently, LANDSAT 8 OLI satellite images with level 2 processing (L2SR), which present top-of-atmosphere reflectance (TOA) data, are generated from level 1 collection (L1C) images [4].

2.2.2. Sensor TIRS. Corresponds to the thermal imaging instrument that generates thermal images characterized by a spatial resolution of 100m and measures in the long-wave infrared spectrum to study surface temperature variation, heat and moisture transfer process in agriculture, water management, etc. [18].

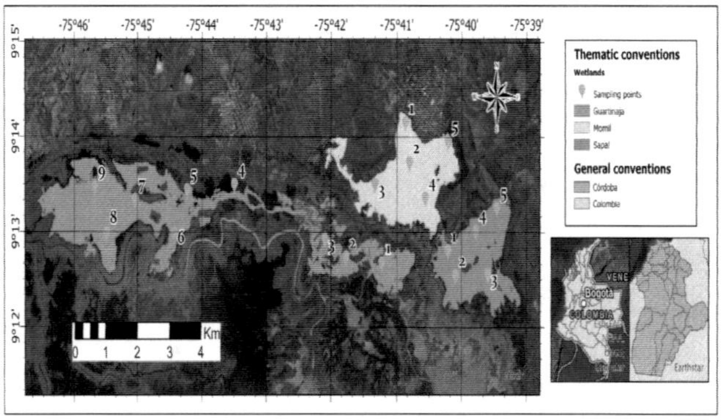

Figure 1. Study area general location.

2.3. Primary data organization of water quality measurements

Water quality parameter data were taken from monitoring campaigns conducted in Guartinaja, Momil, and Sapal wetlands, carried out on February 5 and 15, 2021 as part of the University of Córdoba Environmental Engineering program's internal project "Hydrodynamic and water quality monitoring strategy based on the combination of mathematical models and reflective images applied to two wetland ecosystems in the department of Córdoba". Environmental parameters were measured at 19 sampling points as shown in Figure 1.

2.4. Satellite image processing

The LANDSAT 8 satellite spectral images from the OLI sensor and bands 10 and 11 from the TIRS sensor were downloaded from the USGS platform for February 3 and January 1, 2021. For the surface delimitation of the Guartinaja, Momil, and Sapal wetlands, the normalized differential water index (NDWI) defined by Equation 1 in ArcGIS Pro software was applied, using the GREEN (visible green) and NIR (near infrared) spectral bands, because, beyond the latter, water does not reflect considerable light [15].

$$NDWI = \frac{(GREENNIR)}{(GREEN + NIR)} \tag{1}$$

Journal of Physics: Conference Series **2475** (2023) 012007 doi:10.1088/1742-6596/2475/1/012007

2.5. Statistical evaluation of dependent and independent variables

The band reflectance values used for each pixel where the sampling points were located were extracted from the reference satellite image (February 3), for which a rescaling was done using equation 2, where DNout values are from 0 to 1 [25].

$$DN_{out} = Min_{out} + \frac{(DN_{in} - Min_{in})X(Max_{out} - Min_{out})}{Max_{in} - Min_{in}} \quad (2)$$

In the thermal bands' case (10 and 11), the brightness temperature was calculated; for this purpose, the digital levels (ND) were converted to radiances using equation 3, where L corresponds to the spectral radiance received by the sensor in a band in $Wm^{-2}sr^{-1}\mu m^{-1}$, $a_0 and a_1$ are the calibration coefficients obtained from the metadata for each band, and ND is the pixel's digital level in that band [25].

$$L = a_0 + a_1 \cdot ND \quad (3)$$

Once the radiance (L) is calculated by Eq. 3, the brightness temperature (T_B) according to the Planck equation inversion, is estimated by Eq. 4, considering that k_1 (in $Wm^{-2}sr^{-1}\mu m^{-1)}$ and k_2 (in degrees K) correspond to each band constants calibration.

$$T_B = \frac{K_2}{\ln(\frac{K_1}{L}) + 1} \quad (4)$$

For model equations' construction by multiple linear regression (MLR), the Shapiro-Wilk test was applied to determine the normal distribution, and data transformation was required by applying mathematical adjustments.

2.6. Statistical model estimation for each water quality parameter

The Stepwise Regression method was used with MatLab R2022 software to determine the statistical models for the parameters from the database generated with reflectance values and parameters in-situ with a normal data distribution; in such a way the models were chosen according to the statistical criteria corresponding to the root mean square error (RMSE) which must be less than 0.5, the determination coefficient (R2) greater than 0.6 and p-value less than 0.05 [20].

2.7. Statistical models digitalization

The model equations were entered into the raster calculator of the QGIS software using the bands of the Landsat 8 images, firstly for the field sampling date of February 5, 2021, as one of the criteria when choosing the model with the largest area of the body of water to cover and that was best associated with the natural behavior of the parameters studied, it was also found that the estimated data on the sampling points had a good correlation concerning the in situ data, which are represented in the graphs of correlation (Figure 2). After this, the models were digitized using the respective bands of the date of February 2, 2021.

3. Results and discussion

3.1. Field data

The coordinates and each parameter's value previously organized by wetlands for February in CSV format were uploaded to the ArcGIS Pro program and georeferenced in the form of points; the wetlands shape during the dry season, extracted from LANDSAT 8 satellite images dated February 3, 2021, were also attached through the spectral index NDWI.

Journal of Physics: Conference Series **2475** (2023) 012007 doi:10.1088/1742-6596/2475/1/012007

3.2. Models for estimating water quality parameters

Table 1 presents obtained models for each water quality parameter; likewise, Figure 2 shows the correlation between the in-situ data and the estimated data for each parameter. These models are the result of the best possible relationship between the in-situ values of February 5, 2021, and the reflectance values of the spectral and thermal images of the LANDSAT 8 satellite of February 3, 2021.

For dissolved oxygen, the square root of the in-situ data was taken for the significant reduction of the root mean square error (RMSE) to 0.141; on the other hand, the R^2 for this parameter indicates a precision higher than 80% and a significance lower than 0.5. The spectral bands that best correlate with the estimated variable, in this case, are Blue (band 2) and Red (band 4), and within the thermal bands, the TIR 2 band (band 11); while for the pH model equation, which meets the statistical criteria, the related spectral bands are Red (band 4) and SWIR 2 (band 6) in the range 0.452 to 1.651 μm,which coincide with the bands used in investigations carried out at El Guájaro reservoir, Atlántico by [10]; in La Paz bay, Mexico by [28]; at Colorada Bay Beach, Sinaloa, Mexico [7]; at Owabi reservoir, Kumasi – Dhama [9], where the most repeated is the LANDSAT 8 Red band (band 4).

Finally, the temperature and turbidity parameters' model equations, in addition to meeting the statistical criteria, integrate the bands in the spectrum from 0.533 to 1.651μm wavelength, going from the green band (band 3) to the SWIR 1 band (band 6), additionally, the temperature model integrates the thermal band TIR 2 (band 11) as used by [10] in their respective temperature models.

Table 1. Linear regression models for dissolved oxygen estimation.

Linear regression model	RMSE	R_2	p-value
$\sqrt{DO} = 3.940628.099(B4)^2 0.32123(\frac{B4}{B2})1.08(10)^9 \cdot(B11) + 32.083(\frac{B4^3}{B2})$	0.141	0.804	7.38e-05
$pH = 10.454 + 18.352(B4)3.1284(\frac{B4B5}{B4+B5})$ $15.115\ (\frac{(B6}{B2})\ \ 10.2(\frac{B4 \cdot B2}{B3})$ $+\ 26.32\ (\frac{B6}{B2})2$	0.432	0.660	0.00874
$Temp = 42.602 + 46.462(B3) + 682(B5)^2 2.0284(\frac{B3}{B6})$ $1.157(10)^{10} \cdot (B11)10130(B3) \cdot(B5)^2 + 1.3373(10)^9 \cdot \frac{B11 \cdot B3}{B6}$	0.576	0.794	0.00144
$LogTurb = 0.19552 + 5.3989(B4) + 0.38474(\frac{B6}{B5})$	0.207	0.634	0.000322

DO: dissolved oxygen; temp: temperature; LogTurb: Logarithm of turbidity; B2: Blue band; B3: Green band; B4: Red band; B5: NIR band; B6: SWIR1 band; B7: SWIR2 band; B11: TIRS2 band

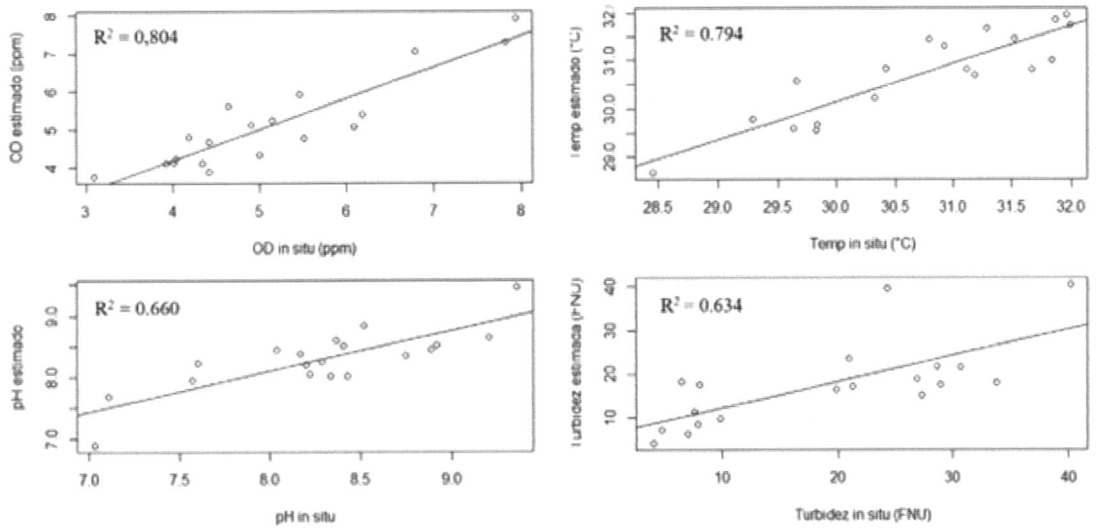

Figure 2. Scatterplot and correlation between in situ measured and estimated values of the studied water quality parameters.

3.3. Digital models

3.3.1. Dissolved Oxygen. It is the amount of gaseous oxygen found in the water and is an important indicator of water quality, especially for aquatic life [6]. Figure 3 shows dissolved oxygen distribution in the water body, it can be observed a very evident variation between the two dates studied in the range of 3.1 and 8.2 ppm. In Figure 3a, the parameter's homogeneous distribution and concentration are found specifically in the Guartinaja wetland, while in Momil and Sapal there are high and low concentrations in the areas near each water body bank. This uniformity DO distribution also occurs in Figure 3b except for the Guartinaja wetland, because, to the west of the wetland, there is an increase in the dissolved oxygen concentration where there is a greater variation in the entire water body.

Since these are dry season dates, aerobic conditions can increase and favor fish species diversity, but the temperature is a factor against this, since it influences the oxygen capacity to dissolve, so an inverse relationship can be established, as shown in Figure 3 and Figure 5 [11]. On the other hand, the dissolved oxygen (DO) remains above the desirable minimum in all the swamps, except for the date of February 5 in the center-east zone and the extreme west of the Guartinaja swamp; in the south and center of the Momil swamp and the western zone of the Sapal swamp.

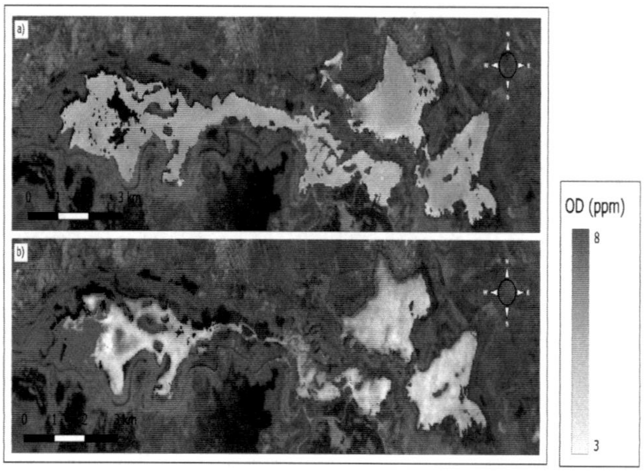

Figure 3. Dissolved Oxygen Modeling for January 02 (a) and February 03 (b), 2021.

3.3.2. pH. It is defined as the negative logarithm of the concentration of hydrogen ions in a solution and expresses it is associated with the degree of acidity and alkalinity of the solution [9]. Considering Figure 4b, in the Guartinaja wetland western area, high pH values were maintained, while in the Momil and Sapal wetlands, a tendency towards neutrality can be observed over a large part of the surface. In Figure 4a the trend from basic to neutral values was much more marked than at the later date.

The pH is proportional to the temperature (Figure 5). In his study, Hernandez found 2018 high pH values during months with cold seasons (January-June) and low values in the months of August-November supported by the fact that temperature affects the water's dissociation constant and produces changes in the hydronium and hydroxyl ions' relative concentration, also, a high relationship between the two parameters was found in a study of Baja California west coast [11]. In the case of pH, for both dates, some areas do not comply with the range of 6.5 to 9, which tends to the basic pH state above 9 but starts to decrease on February 5.

3.3.3. Temperature. Temperature measures the water molecules' average kinetic energy and represents one of the most influential factors in the dynamics of several water quality parameters such as dissolved oxygen and pH because it promotes microorganisms' growth and proliferation, as well as the aquatic vegetation photosynthesis process [23]. The results indicate that the highest temperature values are observed in Figure 5b, few are low-temperature zones, more specifically the area near the Guartinaja

wetland western edge. In contrast, in Figure 5a there was a decrease in temperature, with a higher percentage of areas with low temperatures near 23°C being evident.

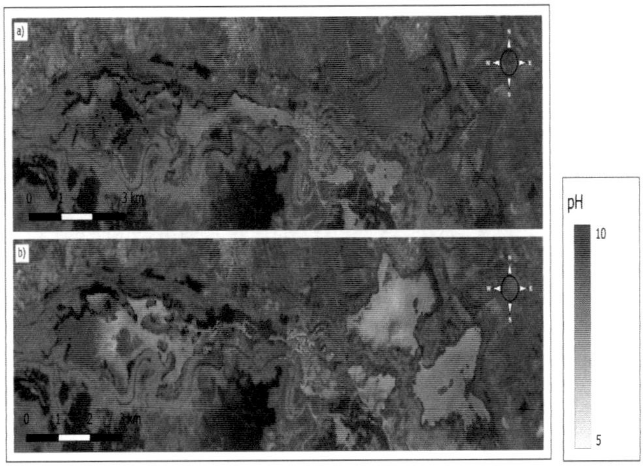

Figure 4. pH modeling for January 02 (a) and February 03 (b), 2021.

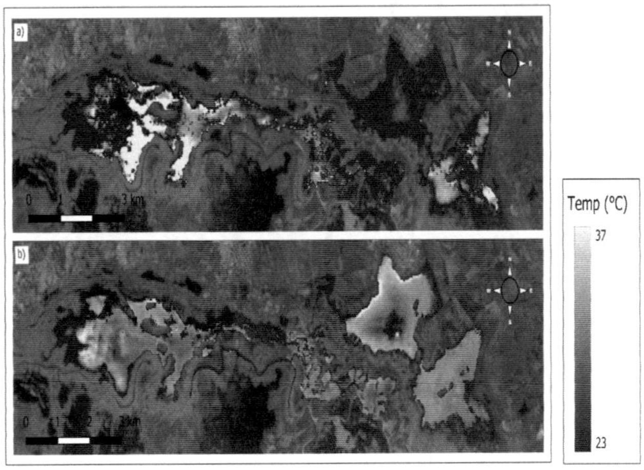

Figure 5. Temperature modeling for January 02 (a) and February 03 (b), 2021.

3.4. Turbidity.

Turbidity. It is associated with the content of suspended particles in the water and is therefore a fundamental physical indicator for evaluating water quality. The highest turbidity values were reached in the Guartinaja wetland western part for the February date in Figure 6b. Momil wetland shores also have high turbidity values, around 65.5 FNU, which decreased over time as shown in Figure 6. It is noteworthy in Figure 6a that Momil wetland shores also have high turbidity values, around 65.5 FNU, which decreased over time concerning the date of January 3 (figure 6a), while for Momil it will occur slightly the following month. The increase in turbidity in edge zones may be due especially to sediments deposition that the tributary drags towards the wetland complex.

Regarding the regulations, turbidity did not meet the minimum conditions for human consumption in the swamps for both dates, since it is well above the limit of 2 NTU, reaching values of up to 65.5 NTU.

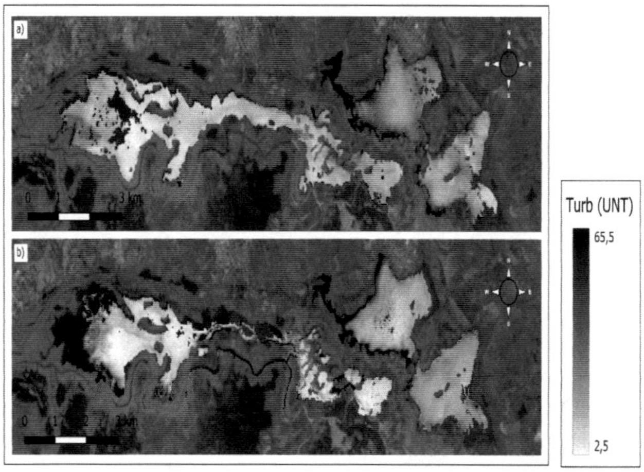

Figure 6. Turbidity modeling for January 02 (a) and February 03 (b), 2021.

4. Conclusions

The combination of remote sensing and multiple linear regression methods for water quality parameters estimation through statistical models is a useful monitoring method, facilitating the wetlands' environmental, social, and economic management. The use of Remote Sensing with LANDSAT 8 satellite images for the estimation of mathematical models of water quality parameters represented an advantage since it does not require preprocessing for level 2 images because they are integrated and include thermal bands that increase the bands' numbers that correlate with important parameters such as temperature and dissolved oxygen.

The obtained models presented a good correlation between the in-situ values and those estimated for each water quality parameter for the field data collection date (Figure 2), and consequently, satisfactory results were obtained for the parameters' digital models studied on both dates. Cloudy conditions covering the study area should be avoided because they limit the models' application, preventing results generation.

The models developed from these satellite images help to understand the different physical, chemical, and biological processes occurring in the study area, thus preventing negative anthropogenic inputs. The implementation of these technologies is essential for water quality assessments along with national and international regulation according to the use either for human consumption or for aquatic life protection. The assessment of water quality through the national regulation for human consumption by the RAS and resolution 2115, the swamps do not have sufficient conditions to be used for human consumption since the water did not meet the oxygen parameters dissolved, pH, and turbidity in edge zones.

References

[1] Diego Alonso. *Combinación de bandas en imágenes de satélite Landsat y Sentinel.* MappingGIS, May 2019. URL: https://mappinggis.com/2019/05/combinaciones-de-bandas-en-imagenes-de-satelite-landsat-y-sentinel/.

[2] *Aplicación de datos del mes: Monitoreo de la calidad del agua — UN-SPIDER Knowledge Portal.* www.un-spider.org. URL: https://www.un-spider.org/es/enlaces-y-recursos/fuentos-de-datos/daotm-monitoreo-calidad-agua (visited on 01/15/2023).

[3] Anderson Berdugo Muñoz. "Análisis multiespectral de imágenes satelitales para evaluar variaciones espacio - temporales en la calidad del agua del embalse el Guájaro, Atlántico." In: *repositorio.cuc.edu.co* (Sept. 2016). URL: http://hdl.handle.net/11323/325 (visited on 01/15/2023).

[4] *Cálculo de reflectancia en imágenes Landsat OLI-8, sobre la región central de Honduras, mediante software libre SEXTANTE. - PDF Descargar libre.* docplayer.es. URL: https://docplayer.es/56750831-Calculo-de-reflectancia-en-imagenes-landsat-oli-8-sobre-la-region-central-de-honduras-mediante-software-libre-sextante.html (visited on 01/15/2023).

[5] *Estudio Nacional del Agua 2018 - CTA.* Centro de Ciencia y Tecnología de Antioquia - CTA, 2019. URL: https://cta.org.co/biblionet/estudio-nacional-del-agua-2018/ (visited on 01/15/2023).

[6] *Formato Común Hoja Metodológica.* URL: http://www.ideam.gov.co/documents/24155/123679/06-3.18_HM_Promedio_oxigeno_disuelto_3_FI.pdf/0ce75169-1d6a-44fa-bfea-972b6bb32fdd.

[7] Luis Carlos González-Márquez et al. "Estimation of water quality parameters using Landsat 8 images: application to Playa Colorada Bay, Sinaloa, Mexico". In: *Applied Geomatics* 10 (Mar. 2018), pp. 147–158. DOI: 10.1007/s12518-018-0211-9. (Visited on 04/27/2022).

[8] Jerry A. Griffith. "Geographic Techniques and Recent Applications of Remote Sensing to Landscape-Water Quality Studies". In: *Water, Air, and Soil Pollution* 138 (2002), pp. 181–197. DOI: 10.1023/a:1015546915924.

[9] A. K. M. Azad Hossain, Caleb Mathias, and Richard Blanton. "Remote Sensing of Turbidity in the Tennessee River Using Landsat 8 Satellite". In: *Remote Sensing* 13 (Jan. 2021), p. 3785. DOI: 10.3390/rs13183785. URL: https://www.mdpi.com/2072-4292/13/18/3785/htm (visited on 10/07/2021).

[10] Nidal M. Hussein and Mohammed N. Assaf. "Multispectral Remote Sensing Utilization for Monitoring Chlorophyll-a Levels in Inland Water Bodies in Jordan". In: *The Scientific World Journal* 2020 (Aug. 2020), pp. 1–14. DOI: 10.1155/2020/5060969. (Visited on 11/07/2020).

[11] ROSA JUDITH. "Estimación de parámetros ambientales a partir de imágenes Landsat 8 y mediciones in situ, en Bahía de La Paz, BCS". In: *Cibnor.mx* (2018). DOI: http://cibnor.repositorioinstitucional.mx/jspui/handle/1001/1381. URL: http://dspace.cibnor.mx:8080/handle/123456789/2734 (visited on 01/15/2023).

[12] Samy Kharuf-Gutierrez et al. "Análisis de imágenes multiespectrales adquiridas con vehículos aéreos no tripulados". In: *Ingeniería Electrónica, Automática y Comunicaciones* 39 (Aug. 2018), pp. 79–91. URL: http://scielo.sld.cu/scielo.php?script=sci_arttext&pid=S1815-59282018000200007 (visited on 01/12/2023).

[13] . *Landsat 8 — U.S. Geological Survey.* www.usgs.gov. URL: https://www.usgs.gov/landsat-missions/landsat-8.

[14] *Landsat 8 Bands — Landsat Science.* Nov. 2021. URL: https://landsat.gsfc.nasa.gov/satellites/landsat-8/landsat-8-bands/.

[15] Juan Manuel et al. *EDICIÓN CARTOGRÁFICA Grupo de Redes IDEAM.* URL: http://corpouraba.gov.co/wp-content/uploads/2.-PROTOCOLO_MONITOREO_AGUA_IDEAM.pdf (visited on 01/15/2023).

[16] *Ministerio de Desarrollo Económico 2000 Resolución 1096.* Nov. 2000. URL: http://www.oas.org/dsd/environmentlaw/waterlaw/documents/Colombia-Resolucion_1.096_(2000).pdf.

[17] *MinSalud MinAmbiente 2007 Resolución 2115.* June 2007. URL: https://laboratoriodeanalisis.lasalle.edu.co/wcm/connect/LIAC/d951c109-a227-44a3-8a42-1d1f87db2b43/Resoluci%5C%C3%5C%B3n_2115-2007.pdf?MOD=AJPERES&CVID=lMoOSFe.

[18] Matthew Montanaro, Raviv Levy, and Brian Markham. "On-Orbit Radiometric Performance of the Landsat 8 Thermal Infrared Sensor". In: *Remote Sensing* 6 (Nov. 2014), pp. 11753–11769. DOI: 10.3390/rs61211753. (Visited on 03/07/2021).

[19] A.A. Obaid et al. "Assessing the utility of using current generation high-resolution satellites (Sentinel 2 and Landsat 8) to monitor large water supply dam in South Africa". In: *Remote Sensing Applications: Society and Environment* 22 (Apr. 2021), p. 100521. DOI: 10.1016/j.rsase.2021.100521. (Visited on 08/28/2021).

[20] *Regresión Lineal Múltiple.* URL: http://eio.usc.es/eipc1/BASE/BASEMASTER/FORMULARIOS-PHP-DPTO/MATERIALES/Mat_50140129_RegresionMultiple.pdf.

[21] Dennis Reuter et al. "The Thermal Infrared Sensor (TIRS) on Landsat 8: Design Overview and Pre-Launch Characterization". In: *Remote Sensing* 7 (Jan. 2015), pp. 1135–1153. DOI: 10.3390/rs70101135. (Visited on 06/26/2022).

[22] Burgos Doria Rosalía. "Significado de valor cultural, natural y ambiental del humedal Ciénaga Grande del Bajo Sinú para los habitantes de la Vereda Caño Viejo (Lorica – Córdoba – Colombia)". In: (May 2015). URL: https://ridum.umanizales.edu.co/xmlui/handle/20.500.12746/2091 (visited on 01/15/2023).

[23] Neelima Satyam. *Chapter 5 - Numerical modeling of biocemented soil behavior.* Ed. by Pijush Samui et al. ScienceDirect, Jan. 2021. URL: https://www.sciencedirect.com/science/article/pii/B9780128212059000150.

[24] Rodrigo Takashi. "DISEÑO DE MODELOS DE CALIDAD DEL AGUA MEDIANTE EL USO DE LA PERCEPCIÓN REMOTA". In: *52.100* (2021). DOI: http://132.248.52.100:8080/xmlui/handle/132.248.52.100/4562. URL: http://132.248.52.100:8080/xmlui/handle/132.248.52.100/4562 (visited on 01/15/2023).

[25] Ajay Kumar Taloor, Drinder Singh Manhas, and Girish Chandra Kothyari. "Retrieval of land surface temperature, normalized difference moisture index, normalized difference water index of the Ravi basin using Landsat data". In: *Applied Computing and Geosciences* 9 (Mar. 2021), p. 100051. DOI: 10.1016/j.acags.2020.100051. (Visited on 04/14/2021).

[26] F. Torres-Bejarano et al. "Water quality assessment in a wetland complex using Sentinel 2 satellite images". In: *International Journal of Environmental Science and Technology* 18 (Oct. 2020), pp. 2345–2356. DOI: 10.1007/s13762-020-02988-3. (Visited on 06/02/2022).

[27] Rubén Darío SEPULVEDA Vargas, María Alejandra TABORDA Caro, and Deivi David FUENTES Doria. "Conflictos por el agua y resiliencia comunitaria en el Bajo Sinú: evidencias de disputas y alternatividad. / Water conflicts and community resilience in Lower Sinu: Evidence of disputes and alternativity." In: *Utopía y Praxis Latinoamericana* 25 (July 2020), pp. 105–124. URL: https://produccioncientificaluz.org/index.php/utopia/article/view/32645 (visited on 01/15/2023).

[28] Jaris Veneros et al. "Aplicación de sensores remotos para el análisis de cobertura vegetal y cuerpos de agua". In: *Idesia (Arica)* 38 (Dec. 2020), pp. 99–107. DOI: 10.4067/s0718-34292020000400099. (Visited on 08/02/2021).

[29] X. Xiao et al. *Evaluation method of water quality for river based on multi-spectral remote sensing data.* www.int-arch-photogramm-remote-sens-spatial-inf-sci.net, Apr. 2015. DOI: 10.5194/isprsarchives-XL-7-W3-1517-2015. URL: https://www.int-arch-photogramm-remote-sens-spatial-inf-sci.net/XL-7-W3/1517/2015/ (visited on 01/15/2023).

Cost Model for CubeSat Projects in Mexico

E N Jiménez Navarro[1], A Solís Santomé[2],D A Padilla Pérez[3], F A Domínguez Pacheco[4], C J Sosa Paz[1] and C Silva Sánchez[5]

[1]National Polytechnic Institute, SEPI-ESIME, Zacatenco. Program in Systems Engineering. Professional Unit, Av. Luis Enrique Erro, Adolfo López Mateos S/N, Zacatenco, Gustavo A. Madero, 07738 Mexico City.

[2]National Polytechnic Institute, Linkage and Regional Development Center, Cancun Unit. Business Center, Blvd. Kukulcan Km 12.5-Tower A, Hotel Zone, 77500 Cancún, Q.R.

[3]National Polytechnic Institute, CDA, SARACS Research Group. Professional Unit, Av. Luis Enrique Erro, Adolfo López Mateos S/N, Zacatenco, Gustavo A. Madero, 07738 Mexico City.

[4]National Polytechnic Institute, SEPI-ESIME, Zacatenco. Program in Systems Engineering-Biophysical Systems in Agriculture and Food. Professional Unit, Av. Luis Enrique Erro, Adolfo López Mateos S/N, Zacatenco, Gustavo A. Madero, 07738 Mexico City.

[5]National Polytechnic Institute, CIDETEC-IPN, Deputy director of technological innovation. Av. Juan de Dios Bátiz S/N, Nueva Industrial Vallejo, 07700 Gustavo A. Madero, CDMX.

E-mail: `ejimenezn1200@alumno.ipn.mx`

Abstract. This work focused on designing a cost estimation model in the area of space engineering for the development of nanosatellite projects in Mexico. See employed the use of methodologies such as Hall and Jenkins, software engineering cost estimating methodology, as well as project management methodology. Similarly, the COSYSMO methodology and parametric cost methodology were incorporated to guide this research towards the space sector. The estimation results are based on satellite engineering equations and forecasts of the inflation factor. Finally, a graphical user interface named SACANI was obtained. Allowing the estimating a nanosatellite project by "integration of components" or by "research and development". The first asks the user for the components by subsystem through drop-down lists, giving a breakdown of the estimated costs for the CubeSat, development phases and ground station, according to the data provided by the interested party. The second model is more robust because it requests the type of project, the category, the amount of CubeSat required, the year to estimate the cost, the mass of the payload, among other data. Similarly, both models yield a previous result of the estimate made according to the data entered by the project estimator.

Keywords: CubeSat; Model; Cost; Lifecycle; Methodology.

1. Introduction

Communication flows within our bodies, thus having access to more information through multiple channels improves life [24]. Throughout history, human beings have sought to communicate with their peers in a fast and efficient way, obtaining results such as telecommunications [16]. Communication technologies are divided into wired and wireless communications, the former use a physical medium to send a signal, while wireless ones travel in the form of waves [5]. In this order of ideas, artificial satellites are found in wireless communications, defined as a spacecraft created by man and launched into space to orbit around a celestial body. The first satellite launched was Sputnik-1 in 1957 by the Union of Soviet Socialist Republics (USSR) [21]. On the Mexican side, it was not until 1982 that it acquired its first own satellite system, which was known as the "Morelos System" made up of the Morelos I and Morelos II satellites. For 1993 and 1994, the Morelos system was replaced with the "Solidarity Satellite System" made up of the Solidaridad I and Solidaridad II satellites, satellites acquired by the Mexican government [3]. In 2012 the Bicentennial was launched and at the end of 2015 the Morelos III was launched [14]. From the foregoing, it is clear that of the satellites that Mexico has none has been built by Mexican technology. Currently, there are different types of satellites classified by their weight and dimensions. Technological progress has made it possible to miniaturize electronic components, achieving the development of small satellites. It was in 1999 that the CubeSat project started and a new class of picosatellites "the CubeSat standard" was developed. This new class of satellites allowed companies and universities to have access to the financing and development of the CubeSat. Within the cost estimate, elements such as the satellite subsystems, launcher, earth station, among others, are considered. The control of the cost estimation allows to determine the sale price and profit. If a good estimate is not made, it can result in financial losses. This research focuses on cost estimation for the development of nanosatellite projects under the CubeSat standard, achieving efficient investment planning of the project budget [17] and [6]. In section 2 talks about the different cost models that were reviewed, as well as the methodology used for the development of this research. Section 3 presents costs associated with components and development phases of the CubeSat. Section 4 shows the analysis of the data on the inflation factor, then some of the equations used in estimating the cost of a CubeSat are indicated. Section 5 is about results, where the cost model is observed in a graphical user interface. Finally, in section 6 aspects of the development and scope of the research are discussed.

2. Materials and methods

Cost Methodologies use specific parameters and employ models that have existed for a long time [10]. Cost estimation allows project managers to assess project feasibility, cost risk, and possible alternative designs, allowing stakeholders to better understand the mission, requirements, and development of projects [22] and [10].

2.1. Experience-Based Cost Model

This type of cost estimation uses expert judgment as the primary input. The cost prediction is based on the experience of the experts. The participation of several estimators is required to give their opinion of the effort required in the development of the project. The problem with this model is evident in the mistakes of the experts' judgment. Then, there is the Constructive Cost model (COCOMO). Through basic equations, it allows estimating the number of individuals per month to complete the work in that period of the project's life. In addition, it calculates the months required to complete the project [15]. Also, as a reference to this work, it is revised the Construction systems engineering cost model (COSYSMO). Created by Ricardo Valerdi when he was at the Software Engineering Center at the University of Southern California [39]. This model calculates the number of person-months, engineering resources on the project in both hardware and software. It calculates the effort based on the size of the system to later be adjusted according to a series of factors related to the engineering of the system [39] and [29].

The variety of projects has allowed the development of cost estimation models that adapt to the needs of each project. To mention a few more, there is the PRICE-H space hardware commercial cost model, which is frequently used in the aerospace sector [19]. Air Force Cost Model (NAFCOM) was developed by Marshall at the Space Flight Center (MSFC) and is used for manned spacecraft, uncrewed spacecraft, and launch vehicles in their early development phases [40] and [20]. Advanced Mission Cost Model (AMCM) developed by the Office of Exploration Programs at Johnson Space Center, intended to be a long-term cost forecasting tool [35], [23], [31], [9] and [9].

2.2. Theories involved in the cost estimation model

Cost estimation is based on different theories that allow synergy and homeostatic balance for the development of research.

2.2.1. Decision theory It is a mathematical theory that deals with choices between possibilities [36]. Choices between possibilities or decision making is an intermediate phase between formal and informal model theories. Some factors that influence decision making are: risk and how to deal with it, human factors, pressure and stress, costs, among others [28]and [27].

Also, the Analytic Hierarchy Process technique (AHP), it is based on the analyst choosing a comparison value using a scoring scale [12]. These, will be used within the cost estimation process.

The smoothing method will be used in inflation factor data to remove noise from the data to help reveal important features such as trend, seasonality. there are also adaptive models that allow corrections in the parameters as the forecasts are generated [34].

Then, Brown's method will be used in a branch of exponential smoothing methods, it is applied when the data pattern is not linear, it can be adjusted to quadratic, cubic or higher order functions [34]. Then, the validation of the forecast is required, through the goodness of fit and Theil's U test. Later, will be used the goodness of fit as a measure of how well the selected model reproduces the original data. While Theil's U allows a relative comparison between formal forecasting methods and intuitive or elementary methods of judgment [34].

Afterwards, a graphical interface will be made with the cost estimation model. The graphic interface implies the presence of a computer monitor or screen constituted by a series of menus and icons that represent the options that the user can take within the system [33], [26], [25], [38] and [30].

2.3. Methodology

The methodology explains what is related to the type of design or method, as well as the procedure carried out for the study [13] and [41]. This work is based on the interpretation of real cost methodologies analyzed from various works and which for this research are classified into two categories called: general methodologies and specific methodologies, as shown in Figure 1.

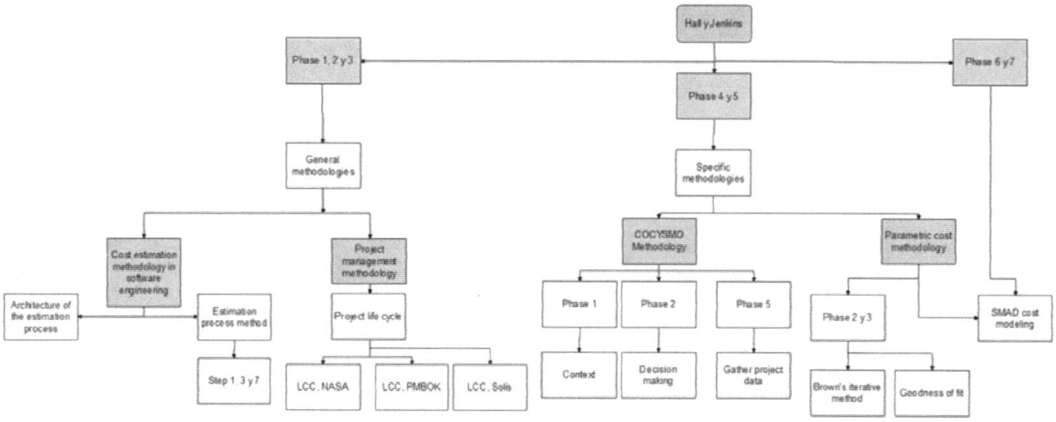

Figure 1. Methodology of this research work, 2020

The general methodologies were implemented to structure the development of this work, while the specific methodologies focus on the space sector. This is why the COSYSMO methodology and parametric cost methodology are used.

3. Costs associated with the estimation model

Given that there is no cost model for nanosatellite projects in Mexico, this model should estimate the cost of the necessary personnel for the different stages of the project, as well as the legal regulations that imply developing a CubeSat in Mexico.

The ISO-15288 standard [2] mentions that any system of interest, regardless of its dimensions, requires a life cycle. Which will be segmented by phases or stages to ensure satisfactory progress. Likewise, there are systems that depend on life cycles that last for decades, such as airplanes, satellites, ships, among others. For this research, the life cycle of [37] is implemented.

The time period is based on the paper from [7]and [4]. Properly adjusting the phases and activities to carry them out in a period of between two and three years to develop each of the CubeSat activities. The time varies depending on the need for the CubeSat payload and the materials to be implemented.

Missions are divided into ranges of duration and each of them involves funding and staffing levels[18].

Thus, the fund for these three phases is 2,418,818 MXN$, according to [8] and [32]. Using the analytical ranking technique, the appropriate percentage for phases 1,2,3 will be estimated. All three phases are evaluated against key criteria. Subsequently, the three criteria will be evaluated by a superior criterion called the "Greatest benefit".

After applying the ranking technique, the weights of the first three criteria are in a matrix and the fourth criterion will be the vector. The matrix is then multiplied by the vector to obtain the final hierarchy.

Now, if 100% of the cost of these three phases is 2,418,818 MXN$, then each of the first three phases is assigned its required cost estimate percentage. Taking into account that phase 1 corresponds to 9.86% of the cost, phase 2 corresponds to 22.703%, while phase 3 corresponds to 67.4%. It is important that the measurements of the designed CubeSat fall within the tolerances of the CubeSat standard.

The cost associated with phase 4 is based on the document [1], in which it says that it corresponds to 2% of the total system, which corresponds to phase 6, "Manufacture and Acquisition of Components", where the total cost of the system, as will be seen later, is made up of space components, instruments and a ground station (it does not include the cost of the launch vehicle), Phase 6 has a total estimated cost of 7,744,597.75 MXN$, so if the estimated cost is of 2% in Phase 4 will correspond to 154,892 MXN$.

Phase 5, Preliminary Design, is required to design the satellite structure and each of the corresponding CubeSat subsystems [37]. At the same time, the license process of the signal that it will occupy must be taken into account. Within the satellite networks not subject to coordination, the ITU does not charge for the amateur radio frequency bands. However, the procedures do carry a cost. The cost assigned to the procedures is shown in Table 1.

Table 1. Cost of regulatory procedures, 2020.

Procedure	Cost (USD$)	Cost (MXN$)
API Registration cost with exception: amateur radio service (ITU council agreement 482).	534.48	12,778.79
Cost of notification and registration Master International Frequency Register (MIFR) with exception: amateur radio service (International Telecommunication Union (ITU) council agreement 482).	7,225.81	172,759.44
Coordination request cost. Depending on the number of units.	5,714.87	136,634.78
Total	13,475	322,173

As a rule, the orbit for nanosatellites is considered low circular, at a height of 400 to 650 km. Thus, the cost of this phase is estimated at 322,173 MXN$.
Phase 6 Component Manufacturing and Procurement identifies and selects criteria for testing andevaluating the system. Table 2 shows the cost of some CubeSat components.

Table 2. Cost of Components by Subsystem, 2020.

Components	Weight (g)	Cost(MXN$)
On-board computer subsystem ISIS on-board computer power subsystem	100	421,545.01
Solar panels CubeSat 1U	50	65,254.64
Electric power system IEPS type A	184 ± 5	120,068.55
Control and orientation subsystem		
Magnetorquer board	196	208,814.86
Subsystem of structures and mechanisms		
CubeSat structure of 1 unit	200	65,254.64
Communications subsystem		
CubeSat antenna system for 1U / 3U	77-85	143,560.22
Argus 1000 infrared spectrometer	230	618,614.03
Propulsion subsystem.		385,002.4
1U Nanosatellite Micropropulsion System	300	2,652,459

Based on the costs shown in Table 2, the CubeSat components for 1U are estimated at 5,532,097 MXN$. The cost for the ground station is shown in Table 3.

Table 3. Earth Station Cost, 2020.

	Innovative Solutions In Space (ISIS) Ground Station		
	ISIS ground station for S band	For small satellites in low earth orbit using Very High Frequency (VHF) and Ultra High Frequency (UHF) radio frequencies	ISIS complete ground station for VHF/UHF/S band
Cost (USD$)	72,236.1	67,687.5	93,679.5
Cost (MXN$)	1,710,111.65	1,598,627.81	2,212,500.88

Based on the costs shown in Table 2 and Table 3, taking the "complete ISIS ground station for VHF/ UHF / S band" it is estimated that only the acquisition of components as marked in this phase, the costis 7,744,598 MXN$, this is when the components are already manufactured, their acquisition is made, in the case of manufacturing the design that the user proposes, it results in a variation due to manufacturing, time and materials. Phase 7 Integration, assembly and testing, see the document [32] to know what they consist of. the estimated cost of the launch option ranges from 12,353 dollars per kilogram to 268,095 dollars per kilogram. The launch price varies depending on the orbit, the altitude, the type of launcher. The estimated cost for Phase 7 is obtained by adding the cost of the launcher that the Soyuz-2.1v was selected for the purposes of this work, (The user selects the one that best suits their project) plus the estimated cost of the environmental tests. , to which we have the following: 340,956 MXN$ cost of the launcher plus 250,034 MXN$ estimated cost of environmental tests, giving a total of 590,990 MXN$.

Phase 8 Launch and Operations. This phase focuses on the choice of the launcher and the participants in monitoring the CubeSat once placed in orbit. [37]mentions that the situation in which Mexico finds itself regarding launch technology, it does not have infrastructure or technology for launch vehicles that allow a satellite to be positioned in space". The Mexican Space Agency indicates which area is required to be the operators for the monitoring of the ground station, see Table 4.

Table 4. Monthly pay of the earth station operator, 2020.

Expert	Payment monthly without experience - MXN$
Telecommunications Engineer	13,000
Data analyst	10,000
Systems administrator	15,000
Operations engineer	12,000
Systems Control	12,000
Electronic Engineer	11,000
Aerospace engineers	24,000

Considering that the average life of a CubeSat is approximately two years, the cost of this phase is calculated based on inexperienced operators, considering only one ground station and one employee from each area, as shown in Table 5.

Table 5. Payment of the earth station operator for the life of the CubeSat, 2020.

Expert	Payment to each for 2 years of operation – MXN$	Payment of all operators in 2 years – MXN$
Telecommunications Engineer	312,000	2,328,000
Data analyst	240,000	
Systems administrator	360,000	
Operations engineer	288,000	
Systems Control	288,000	
Electronic Engineer	264,000	
Aerospace engineers	576,000	

Therefore, the estimated cost of Phase 8 is 2,328,000 MXN$.

4. Results of the R&D estimation

Research and Development (R&D), involve the creation of new knowledge, this being a key element for the general progress of society.

Firstly, the weight per standard is required, which is obtained from Cal Poly 2013, and Cubesat, 2018, where it indicates that the CubeSat categories are: 1U, 1.5U, 2U, 3U and 6U, from which its weight is: 1.33 kg , 2 kg, 2.66 kg, 4 kg and 12 kg, respectively. The sum of weights in kilograms is divided into the weight of the satellite bus and the total weight of the CubeSat. The first adds the weight of each subsystem while the second is the sum of the weight of the satellite bus with the payload. Thus obtaining the total weight of the CubeSat.

Now, it is necessary to estimate the inflation factor. Which is calculated according to the historical data that exists in [1] for the year 2024, being estimated using the Brown method with triple smoothing in blue, see Figure 2.

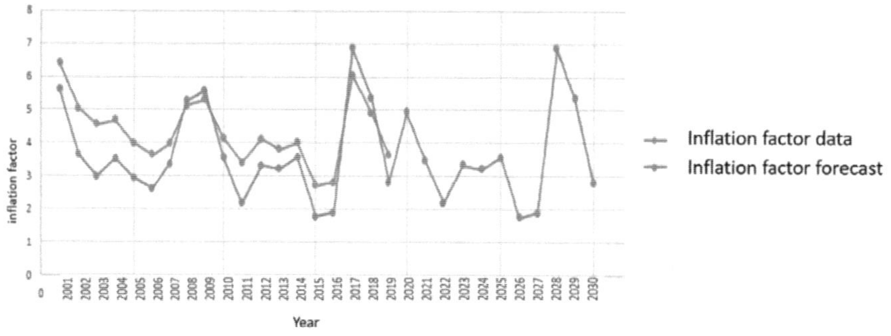

Figure 2. Real data and Forecast, 2020

When applying the goodness of fit to the data obtained by triple smoothing, the results shown in Table 6 are obtained.

Table 6. Goodness of fit of the residuals of the forecast model, 2020.

Goodness of Fit		
Mean Error	ME	0.59464254
Mean of the absolute value of the error	MAE	0.78926656
Error Square Sum	SSE	21.8800615
Mean Squared Error	MSE	0.72933538
Standard Deviation of Errors	SDE	0.86861089
Correlation coefficient		0.93545246
Theil's U		0.01920248

The cost of the dollar in Mexican pesos is considered to be 22.03 MXN$, revised on 06-01-2020, according to [11], the dollar is taken as a reference because it is an international currency.

In addition to the initial data for the estimation, the SMAD considers that for less than 10 units it is recommended to apply a 95% learning curve slope. Between 10 and 50 units a learning curve of 90% and 85% is appropriate for more than 50 units.

To calculate the valor of CubeSat it´s necessary apply "Launch operation" equations.

$$LaunchOperations = 0.061 \cdot (CubeSatTotalWeight = SatelliteBusWeight + Useful\_Cargo1) \quad (1)$$

Then, the program level is:

$$ProgramLevel = 1.963 \cdot (Payload + Bus) \wedge 0.841 \quad (2)$$

Program Level Cost:

$$C.programa = Learningcurve \cdot Programlevel \quad (3)$$

The following equations are increased in Millions of dollars, prior to the end the conversion to Mexican pesos must be made. Now, continue with the investment.

$$I = (Payload + Bus + TechSupportTeam)/1,000,000 \quad (4)$$

Production

$$P = (Lotcost)/1,000,000 \quad (5)$$

Production of the first unit

$$P.firstCubeSatunit = (FirstCubeSatunit)/1,000,000 \quad (6)$$

Launch operations

$$O.L.Cost = (LaunchOperations2)/1,000,000 \quad (7)$$

Program level

$$N.P.Cost = (ProgramLevel)/1,000,000 \quad (8)$$

Therefore, the cost of the CubeSat is calculated with:

$CubeSatcost = ((Investment + Production + Launchoperations + Programlevel) * Inflationfactor)/4$

4.1. Case study
Proposing a hypothetical case study for a remote sensing payload using the R&D cost model. Based on the specifications in Table 7, estimate the cost of a CubeSat in the category of one unit (1U), for the year 2024, under the specifications of the Cal Poly CubeSat Standard.

Table 7. CubeSat data, 2020.

Name	Data
1 CubeSat	1 U
Inflation factor	3.212
Dollar cost	22.03 MXN$
Learning curve	0.95
Payload characteristics	Infrared camera
Opening diameter	1 m
weight of IR	0.45 kg

The result of the cost model was based on the equations of the book "space missions analysis and design", third edition, in the section on cost of the space segment.

Therefore, the cost of the CubeSat is as follows:

$$CubeSatCost = 0.683 Millionsofdollars$$

Which is equivalent to 15,048,143.106 MXN$, (770679.96USD$)

The estimated cost for the development of the phases shown is: 18,677,373,248 MXN$,(956548409.96 USD$).

The cost of the proposed earth station is for "small satellites in low earth orbit using VHF and UHF radio frequencies" shown in the "Manufacturing and acquisition of components" phase. This phase has an associated cost of 7,130,724.81 MXN$, (365195.01 USD$).

In relation to the above, the cost for the development of a CubeSat project is 44,941,865.28 MXN$, (2301665.72 USD$).

It is worth mentioning that each of the aforementioned costs have a 10% interest rate, as well as the adjustment according to the previously forecast inflation factor. In addition, it should be considered in an estimation range that goes from -5% to +10%, according to the PMBOK and the cost study of this research work.

5. Results of the graphical interface
It was developed using the Python programming language, under the SPYDER development environment. The name of the graphical interface is SACANI, this word comes from the Zapotec culture, which translated into Spanish means "the price of the object", which, for this research project, the name SACANI stands for "Sistema de Administración de Costos Aplicado a Nanosatélites de Investigación".

The interface is based on the research developed in the previous sections.

The general structure of the program was made from the system architecture of this work. In Figure 3, the operation of the application is shown in a general way from the time it is initialized until obtaining the estimated cost of the selected project. A data capture and processing process is shown.

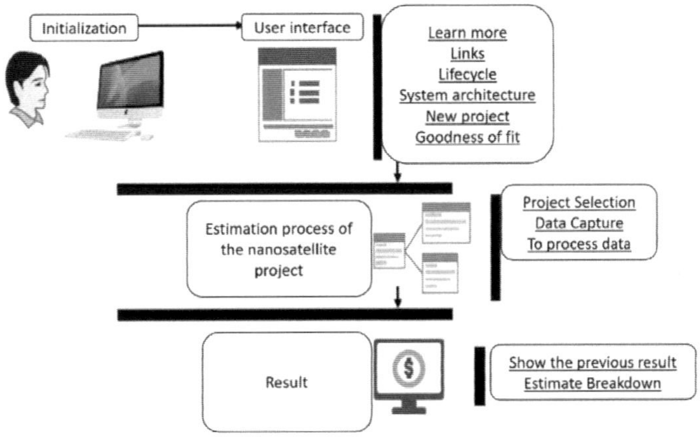

Figure 3. General diagram of the system, 2020.

The first window of the interface is the front page of the program. This consists of the logos of the institution. With this, an initialization code is written that redirects to a window which consists of a button that, when clicked, allows access to the general menu.

When interacting with the interface, there is a button to generate a new project in the main menu, giving rise to a green window where the user is given the option to choose between a project for research and development or for integration of components.

After the interface requests basic data for the estimate, it will display a window with the result and the option of a summary of the estimate.

6. Discussion of results

SACANI allows estimating a nanosatellite project by integrating components or by research and development. The first allows the user to choose the components by subsystem through drop-down lists, giving a breakdown of the estimated costs for the development of your project. The second is more specific because it requires estimator-specific data.

The results obtained in this research document are useful in estimating the development of nanosatellite projects in Mexico, within the CubeSat category of: 1U, 1.5U, 2U, 3U and 6U. On the other hand, the scope of the objectives did not include categories of: 12U, 24U and 36U.

In addition, the estimate of the cost of the project can be calculated within the period of time from 2020 to 2030, this due to the elaboration of a forecast on the annual inflation factor in Mexico, where it was required to know the historical data of the inflation factor. Then analyze the data using Brown's method, obtaining the actual series and the forecast series. Finally, the goodness of fit was used to validate the results of the inflation factor, obtaining a high correlation coefficient in it. To improve the analysis of the inflation factor, it could be analyzed using other types of time series methods, and thus obtain a different forecast, regarding the inflation factor. The forecast does not consider the impact of the "COVID-19" pandemic. As already mentioned, they are data taken from the official INEGI website, which have been analyzed to make the forecast of the inflation factor for the year 2024, as this is a case study in this research work.

The research focuses on the Mexican space sector, this being a reason to place the cost of the different products mentioned here, both in US and Mexican currency.

This cost estimation model did not consider the cost required to move from one point to another, both for the experts and for any component that requires purchase or processing. The present investigation does not contemplate the cost of permits and/or customs procedures, to any country.

References

[1] . *"Índice Nacional de Precios al Consumidor (INPC)"*. 2020. URL: https://www.inegi.org.mx/temas/inpc/ (visited on 09/03/2020).

[2] 14:00-17:00. *ISO/IEC TR 19760:2003*. ISO, 2002. URL: https://www.iso.org/standard/33898.html (visited on 01/15/2023).

[3] Hacia el espacio- AEM. *Historia de los satélites mexicanos*. Hacia el espacio. URL: https://haciaelespacio.aem.gob.mx/revistadigital/articul.php?interior=262 (visited on 01/15/2023).

[4] D. Almeida, F. Mattiello, and F. Sousa. *"Towards Modeling the Concept of Operations for CubeSat-Based Missions"*. 2019.

[5] Pablo Arias Guadarrama. "Clasificación de redes de telecomunicaciones alámbricas e inalámbricas". In: *repository.uaeh.edu.mx* (Nov. 2013). URL: https://repository.uaeh.edu.mx/bitstream/handle/123456789/10884 (visited on 01/15/2023).

[6] J. Arroyave. *Reyes, "Proceso de diseño de una estructura nanosatelital CubeSat"*. Ed. by S. Gutiérrez et al. 2015.

[7] . *Basic Concepts and Processes for First-Time CubeSat Developers NASA CubeSat Launch Initiative*. 2017. URL: https://www.nasa.gov/sites/default/files/atoms/files/nasa_csli_cubesat_101_508.pdf.

[8] Lucinda Berthoud et al. "University CubeSat Project Management for Success". In: *Small Satellite Conference* (Aug. 2019). URL: https://digitalcommons.usu.edu/smallsat/2019/all2019/63/ (visited on 01/15/2023).

[9] *Cost Models - Spacecraft/Vehicle Level Cost Model*. www.globalsecurity.org. URL: https://www.globalsecurity.org/military/intro/reference/calc/SVLCM.htm (visited on 01/15/2023).

[10] J. N. Cruz. *Diseño de una metodología de estimación de costos en ingeniería de software usando el modelo de las capacidades de la madurez del software*. 1998.

[11] *DOF - Diario Oficial de la Federación*. Dof.gob.mx, 2022. URL: https://dof.gob.mx/indicadores.php#gsc.tab=0.

[12] Proceso El et al. *Cabrera, "EL PROCESO DE ANÁLISIS JERÁRQUICO (AHP) Y LA TOMA DE DECISIONES MULTICRITERIO. EJEMPLO DE APLICACIÓN."*. 2008. URL: https://www.redalyc.org/pdf/849/84920503044.pdf.

[13] Miguel Fundación et al. *"Tendencias Epistemológicas de las Investigaciones en los Programas académicos De la división de Estudios para Graduados de la Facultad de ciencias Económicas y sociales De la universidad del Zulia"*. 2014. URL: https://www.redalyc.org/pdf/709/70932556002.pdf.

[14] María Guadalupe et al. *"Diseño de la estructura del micro-satélite Quetzal para la implementación de pruebas de vibraciones*. URL: http://somim.org.mx/memorias/memorias2018/articulos/A1_97.pdf (visited on 01/15/2023).

[15] R. Hana. *"Software Engineering Cost Estimation Using COCOMO II Model"*. Ed. by S. Abeer. 2007. URL: https://www.researchgate.net/publication/315741509_Software_Engineering_Cost_Estimation_Using_COCOMO_II_Model.

[16] C. Hernández Aguilar. *Metodología para la selección de tecnología satelital*. 1999.

[17] J. Herrera Arroyave, S. Santillán Gutiérrez, and P. Sambrano Robledo. *"Proceso de diseño de una estructura nanosatélital CubeSat"*. Ed. by J. Ferrer Pérez and B. Bermúdez Reyes. 2015.

[18] T. Huber. *Guide of Mission Design Activities*. 1991.

[19] Harry Jones. *Estimating the Life Cycle Cost of Space Systems.* 2015. URL: https://ntrs.nasa.gov/api/citations/20160001190/downloads/20160001190.pdf (visited on 01/15/2023).

[20] Bill Keeter. *PCEC – Project Cost Estimating Capability.* NASA, May 2016. URL: https://www.nasa.gov/offices/ocfo/functions/models_tools/PCEC (visited on 01/15/2023).

[21] V.D. Kuznetsov, V.M. Sinelnikov, and S.N. Alpert. "Yakov Alpert: Sputnik-1 and the first satellite ionospheric experiment". In: *Advances in Space Research* 55 (June 2015), pp. 2833–2839. DOI: 10.1016/j.asr.2015.02.033. (Visited on 04/21/2020).

[22] V. L. *"A Survey of Cost Estimating Methodologies for Distributed Spacecraft Missions"*. Ed. by J. Le Moigne and O. L. 2016.

[23] Wiley J. Larson and Linda K. Pranke. "Human spaceflight : mission analysis and design". In: *www.scinapse.io* (Jan. 2007). URL: https://www.scinapse.io/papers/622542928 (visited on 01/15/2023).

[24] James Lull and Eduardo Neiva. *The Language of Life.* Prometheus Books, May 2012.

[25] M. Lurtz. *Powerful Object-Oriented Programming.* 2013.

[26] M. M. Burkhard. *Python GUI Programming.* 2019.

[27] José María and Moreno Jiménez. *EL PROCESO ANALÍTICO JERÁRQUICO (AHP). FUNDAMENTOS, METODOLOGÍA Y APLICACIONES.* 2002. URL: https://users.dcc.uchile.cl/~nbaloian/DSS-DCC/ExplicacionMetodoAHP(ve%5C%20rpaginas11-16).pdf.

[28] E. Mercado Ramírez. *Técnicas para la toma de decisiones :* bibmcgrath.usma.ac.pa, 1991. URL: http://bibmcgrath.usma.ac.pa/library/index.php?title=28459&lang=es&query=@title=Special:GSMSearchPage@process=@autor=MERCADO%5C%20RAMIREZ (visited on 01/15/2023).

[29] M. Mirshams. *"COST AND MASS ESTIMATION MODEL OF SMALL SATELLITES AT SYSTEM DESIGN LEVEL"*. Ed. by M. Samani and A. Darabi. 2005.

[30] A. Moore. *Python GUI programming with tkinter.* 2018.

[31] *NASA Cost Estimating Handbook,* NASA, 2008.

[32] L. P. Parga Beltràn and I. M. Villada Velázquez. *ESTRUCTURA NORMATIVA PARA EL DESARROLLO DEL LABORATORIO DE PICOSATÉLITES CUBESAT-UD DE LA UNIVERSIDAD DISTRITAL FRANCISCO JOSÉ DE CALDAS.* 2016.

[33] J. Paul Mueller. *Beginning Programming with Python for Dummies, John wiley son.* 2014.

[34] B. Paz. *Modelo de Pronósticos.* 2007.

[35] Barney B. Roberts and Dan Bland. "OEXP exploration studies technical report. Volume 3: Special reports, studies, and indepth systems assessments". In: *ntrs.nasa.gov* (Dec. 1988). URL: https://ntrs.nasa.gov/citations/19890006474 (visited on 01/15/2023).

[36] A. A. Sarabia. *Teoría General de Sistemas, México: Fondo de Cultura Económica.* (Visited on 1995).

[37] Arturo Solis Santomé. "Modelado del proceso de diseño de nanosatélites de clase universitaria y análisis de escenarios futuros del sector espacial mexicano". In: *tesis.ipn.mx* (July 2021). URL: https://tesis.ipn.mx/handle/123456789/29085 (visited on 01/15/2023).

[38] J. Sundnes. *SIMULA SPRINGER BRIEFS ON COMPUTING 006.* 2020. URL: https://library.oapen.org/bitstream/handle/20.500.12657/39979/2020_Book_IntroductionToScientificProgra.pdf (visited on 01/15/2023).

[39] Ricardo Valerdi. "The constructive systems engineering cost model (COSYSMO)". In: (Jan. 2005).

[40] Sharon D. Winn and John W. Hamcher. *NASA/Air Force Cost Model: NAFCOM*. ntrs.nasa.gov, Jan. 2002. URL: https://ntrs.nasa.gov/citations/20020048607 (visited on 01/15/2023).

[41] Patricio Yánez Moretta. *Estilos de pensamiento, enfoques epistemológicos y la generación del conocimiento científico Thinking styles, epistemological approaches, and the generation of scientific knowledge*. 2018. URL: https://www.revistaespacios.com/a18v39n51/a18v39n51p18.pdf.

Journal of Physics: Conference Series **2475** (2023) 012009 doi:10.1088/1742-6596/2475/1/012009

Augmented Reality for Interaction with a CubeSat nanosatellite (ARICS)

Jhonatan Fernando Eulopa Hernandez, Emmanuel Cervantes Hernández and Diego Alfredo Padilla Perez

Centro de Desarrollo Aeroespacial, Instituto Politécnico Nacional, CDMX, 06000, México

E-mail: jeulopah@ipn.mx

Abstract. The development of national aerospace technology has aroused interest in different sectors, so there is a long road to transformation where our country must develop the necessary aerospace infrastructure for said technology. The educational sector has been interested in aerospace technology but the use of this type of technology is expensive. That is why, with the support of ICTs, we think about how to solve this problem and this is where augmented reality (AR) comes in, given that today most young people have a cell phone capable of displaying functions of AR through a mobile application. Currently AR in the aerospace industry is used to support the assembly of complex systems such as aircraft or satellites, and perform simulations of the operation of these systems. That is why this project proposes the use of augmented reality so that anyone has access to interaction with a CubeSat-type nanosatellite from their mobile devices.

Keywords: ***CubeSat, Augmented Reality, Mobile apps, Unity, 3D modeling.***

1. Introduction

Augmented Reality integrates 3D models in the field of vision of a user's device, unlike Virtual Reality, it does not replace the real world with a virtual one, but rather complements its environment in a virtual way. This way you don't lose sight of the real world and you can continue to interact with the virtual information that the application shows [1].

Basically four elements are needed to create Augmented Reality which are [2]:

- First of all, you need a device that captures real world images like a cell phone camera or a computer webcam.
- Second, it would be where to display the combination of real and virtual images, in this case the screen of a computer or a mobile phone.
- The next element is a processing device or several of them that work together to interpret information from the real world and generate virtual information and integrate them correctly, these are the processors of a cell phone, the CPU of a computer or a video game console.

- Finally, the last item is the trigger. With which we will be able to visualize the elements developed by 3D elements. There are triggers created by Vuforia software, where images or patterns are used as Qr codes. But also, these triggers can be removed by using Unity AR Core XR and AR Foundation[3] libraries.

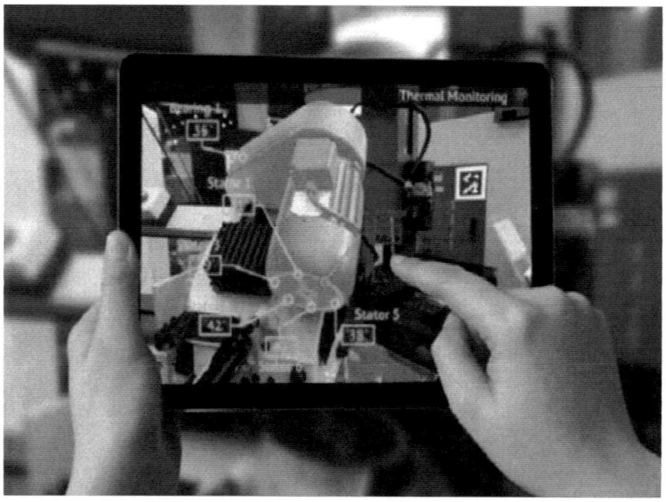

Figure 1. AR architecture example [4]

The successive implementation of new technologies in the classroom, coupled with the growing number of mobile devices, places AR in a notable position. Therefore, AR can be an effective tool in the educational field to change the way students perceive the real world and in this way simplify the complexity of the surrounding world, therefore, from an academic perspective, it provides a greater learning experience.

That is why the multiple applications of AR allow us to develop teaching materials and learning exercises and can be applied in various university disciplines [5].

For the development of this application we begin by choosing which would be the development engine for said application, there are different development engine options but the most popular are Unity and Unreal Engine, so both engines were analyzed to determine the best one to develop the app. , from Which is the easiest to program?, its easy handling, as well as the type of resources offered by the engine for the type of application that would be developed [6].

	Unity	Unreal
Programming	Scripting in C easy to use	Blueprint system programming by nodes / C++
Documentation	Clear and complete, as well as forums and a very large user community	It is not always clear, we only have test projects and there are few users
Graphics	They are not so realistic	Heavier engine, requires more graphics memory
Performance	Light engine, minimal graphics memory	Oriented towards more realistic graphics
APP Weight (Empty Project)	About 20MB	About 40MB

Table 1. Graphics engine [3][7]

After that analysis, it was decided to use Unity, since it uses the C programming language, which is a variant of the C++ language in which you have programming experience and you would only have to learn which libraries Unity uses and as mentioned in the table is easier to use compared to C++ so we could generate more compact scripts and if there is an error we can identify them faster.

Having chosen the ideal graphic engine for the development of the application, we proceeded to choose the type of Augmented Reality that would be used, which can be divided as follows:

- Augmented Reality based on markers
- Augmented reality without markers:
- Location-based AR
- Projection-based AR
- Overlay RA
- contour RA

Knowing the differences would help us choose the necessary AR that satisfies our needs around the type of elements that we would use all of this without the need for external resources other than just the cell phone [8].

- AR based on markers: This type of AR can use different types of markers from a barcode, a QR code, a black and white pattern or a photograph, being captured by the device bed and thus projecting a 3D model .
- AR Without markers: Unlike the previous one, this type of AR positions 3D objects without the need for any marker, partly due to technological advances in mobile devices with their cameras, sensors and Operating Systems.

- Location-based AR: This type of augmented reality uses the sensors of the same device such as GPS, accelerometer and compass in order to obtain the position and location of the user and thus be able to position a 3D model; The most representative example of this type of AR is the game Pokemon Go.
- Projection-based AR: This type of AR focuses on rendering 3D objects within the physical space of the user, allowing the user to move through a certain area where a projector and a tracking camera are placed.
- Overlay-based AR: This AR is used to replace a real object with a manipulable virtual object by displaying additional information that is relevant about the replaced object.
- Contour-based AR: Finally, this AR makes use of special cameras which draw lines in order to facilitate certain situations of daily life, the most common example of this type of AR are the cameras integrated in automobiles that when putting the In reverse, you can see lines that allow safe driving in reverse.

Taking into account the characteristics of the different types of AR, we determined that the best option would be the markerless AR based on location, in this way there is no need to print or project markers or the need for a scale model of the 3D object to be projected. as required by overlay-based AR, the mobile device is simply used [8].

2. Application architecture

ARICS is an application for mobile devices that projects 3D models in a real environment with which we can interact, to achieve this, the space where these models will be placed is scanned, which are selected from a database that is integrated into the interface , once selected we can position and rotate it freely and it can be fixed to the real environment to later display information about the selected model [9].

The architecture of this application is based on an event pattern, being ideal for projects where they can be developed in a modular way and easily maintained.

To apply these events, three states are defined in the application, which are the following:

(i) Main Menu, which will be where the application starts and where we can also view the 3D models fixed in AR. (OnMainMenu)

(ii) Items Menu, where we will find the available 3D models to later select them (OnItemsMenu) and, finally.

(iii) Positioning in AR. (OnARPosition)

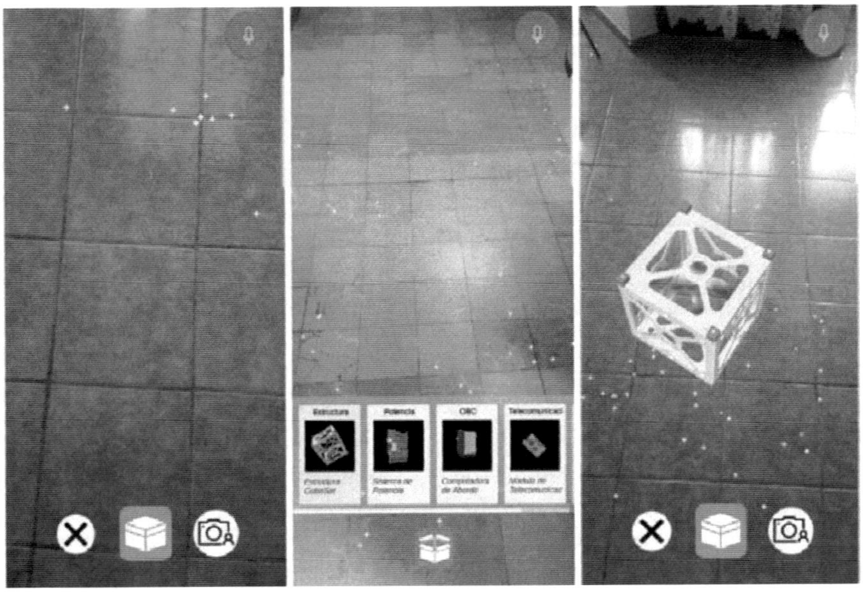

Figure 2. Application States

These three events can be represented with the following pattern: OnMainMenu, OnItemsMenu and OnARPosition Figure (2) respectively, which are included in a script named GameManager in which other scripts can be subscribed to, An example of this is the interface that changes depending on the event that is called.

Once the events are defined, we go to Unity where, before starting with the construction of the application, we must define which platform it will be directed to and thus integrate the fundamental packages for Augmented Reality, which are ARCore XR, which allows compatibility with AR Core Through Unity's cross-platform XR API, the AR Foundation is also added, which allows you to create a specialized work environment for AR development, which complements the ARCore XR package for Android and ARKit XR for iOS.

In Visual Studio we create the structure of our application with the help of the C# programming language. Now having the base architecture of the application, we continue with the design of the interface, development of the 3 states with which it interacts, application of animation to the buttons, with the help of scriptable objects, an efficient Unity tool when saving and ordering information in conjunction with an assets created in C# called item; coupled with a script called DataManager which is responsible for creating a custom button for each 3D model contained in the items.

Using the Blender modeling program, 3D models with the .fbx extension are created and displayed in the software to be integrated into AR once selected in the ItemsMenu.

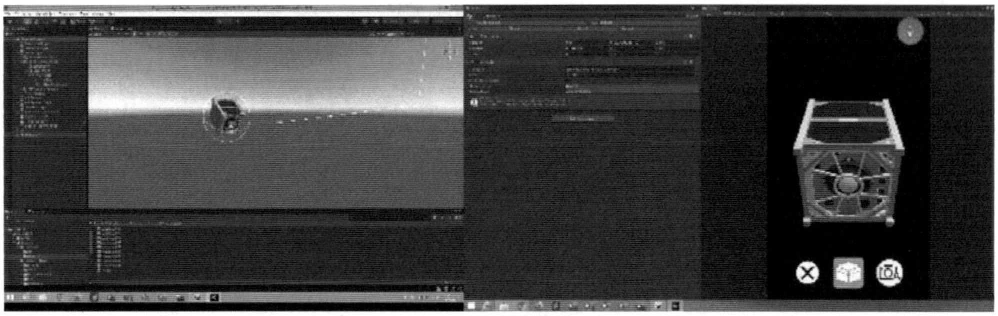

Figure 3. 3D model integrated to Unity

The next step is to create the interaction in AR with the help of a new script called ARInteractionManager where both the positioning of the 3D model and the selection of already positioned models are defined. You have the possibility to delete the models both before and after positioning them on the axis to have a 360° view of the model.

As mentioned before, with this you have the freedom to interact with the 3D models, position them, rotate them and finally fix them in AR so that information about the selected model can be displayed, thus creating a dynamic environment where you learn about the parts of the CubeSat and how it works and the role it has within the CubeSat.

3. Application Usage Results

As a result, we have an application that helps students or any other user to learn about the subsystems of a CubeSat nanosatellite and what is the main function that it performs within the complex system.

It is an intuitive and simple to use application which is ideal for any type of user who wants to learn about aerospace.

Unlike Virtual Reality, which requires controllers and glasses, AR applications do not require these external accessories to interact with the application.

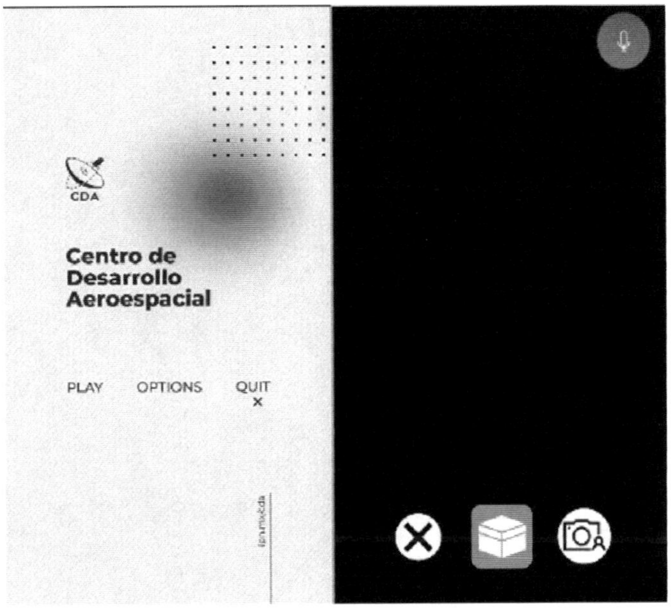

Figure 4. Final Appearance App

4. Conclusions

Augmented Reality and Virtual Reality make use of Information and Communication Technologies to generate applications for education or industry. That is why this AR application of a CubeSat nanosatellite is developed so that more people can have access to technology that shows us the new discoveries of the aerospace industry. All this with low costs since it can be used on any mobile device within our reach.

The application continues to be optimized to provide more functions to the user so that it can compete with applications such as SpaceCraft 3D from (NASA) [10].

References

[1] K. E. C. R. J. O. X. Basogain, M. Olabe, "Realidad aumentada en la educación: una tecnología emergente," *7ª Conferencia Internacional de la Educación y la Formación basada en las Tecnologías. ONLINE*, no. 1, pp. 24–29, 2007.

[2] F. Telefónica, *Realidad Aumentada:una nueva lente para ver el mundo.* International series of monographs on physics, © Editorial Ariel, S.A, 2011.

[3] Unity, "Technologies, u. (s. f.). unity - manual: Manual de unity.." https://docs.unity3d.com/es/530/Manual/UnityManual.html. Accessed: 06.11.2022.

[4] M. logistics trade news, "Tendencias de realidad aumentada para el presente año." https://www.mascontainer.com/tendencias-de-realidad-aumentada-para-el-presente-ano/. Accessed: 10.08.2022.

[5] C. L. M. M. Javier de Pedro Carracedo, "Realidad aumentada: Una alternativa metodológica en la educación primaria nicaragüense," *IEEE-RITA*, vol. 7, no. 2, pp. 102–108, 2012.

[6] Softtek, "¿cuáles son los diferentes tipos de realidad aumentada?." https://softtek.eu/tech-magazine/user-experience/cuales-son-los-diferentes-tipos-de-realidad-aumentada/. Accessed: 06.11.2022.

[7] Unreal, "Unreal engine 5 documentation. (s. f.)." https://docs.unrealengine.com/5.0/en-US/. Accessed: 06.11.2022.

[8] C. Prendes Espinosa, "Realidad aumentada y educación: análisis de experiencias prácticas," *Pixel-Bit. Revista de Medios y Educación*, no. 46, pp. 187–203, 2015.

[9] E. Montoya, "Desarrollo de una aplicación móvil de realidad aumentada. basado en las telecomunicaciones enfocado a la educación," *Repositorio Institucional de la Universidad de Alicante*, 2021.

[10] NASA, "Spacecraft ar and other augmented reality content." https://informal.jpl.nasa.gov/museum/content/spacecraft-3d. Accessed: 25.08.2022.

Journal of Physics: Conference Series **2475** (2023) 012010 doi:10.1088/1742-6596/2475/1/012010

A comparative analysis between Drude and Johnson & Christy for nanometric optical demultiplexer

I Medina[1] and A Villaseñor[2]

[1]Instituto Politécnico Nacional, Centro de Desarrollo Aeroespacial, Mexico City, Mexico
[2]Instituto Politécnico Nacional, Escuela Superior de Física y Matemáticas, Mexico City, Mexico

E-mail: `imedinas@ipn.mx`

Abstract. The constant need to miniaturize technological devices has increased the development of novel optical devices in the nanoscale. A common kind of photonic devices are built using Metal-Insulator-Metal (MIM) interfaces to generate Surface Plasmons Polaritons (SPPs) and process light waves in the optical window. In this paper a 12 channel demultiplexer is proposed using 8 cavities coupled to both sides of the main waveguide. Then the nano demultiplexer is modelled in COMSOL Multiphysics using Drude model and Johnson and Christy parameters for the optical constants. Then we compare the transmission spectrum from 400 to 2200 nm for both models and discuss the differences advantages of each method.

1. Introduction

Due to the constant need of miniaturizing photonic circuits the study and application of SPPs have become more relevant than ever. One of the most interesting characteristics of Surface Plasmons Polaritons is the capacity to overcome the diffraction limit, which allows plasmonic devices to focus and guide light through sub-wavelength structures. Applying these properties is possible to build photonic circuits like waveguides, resonators or demultiplexers in the nanometric scale [8],[2].

For photonic applications the optic properties of a metal are obtained using different physical models, including the Drude model, Interband model or the Johnson and Christy parameters. The appropriate model should be selected according to the material, structure and frequencies of the study.

Optical constants describe the way a material interacts with photons. The real part of the complex refraction index is the classical refraction index n and the complex part represents the extinction index, the first is related to the refraction and the second to the light absorption travelling through the material.

The dielectric constant expresses the capacity of a material to store electric energy or focus electric flux. For solid materials the dielectric constant is more relevant than the refraction index because it also describes the electronical structure of the metal [3].

2. Drude

In 1900 Drude explained the conductive properties of the metals considering that the valence electrons behave like a classical electron gas bounded to a crystalline array of positively ionized atoms. The only force acting the electron gas is the weak bounding to the atoms that prevents the electrons of scaping the surface of the metal. Valence electrons travel freely in the metal until their trajectory is modified due to the collisions with the ionized atoms, after a collision the electron acquires a new random direction and a new speed that is related to the temperature of the place where the collision took place.

When an electric field is applied to the metal, the electrons of the gas travel in the same general direction given by the field, but this path is hampered by collisions with atoms of the array, this resistance is defined by the average time between the collisions of a single electron r. Combining the electromagnetic theory and the kinetic gas theory Drude found an equation that establishes the dielectric constant of a metal in terms of microscopic properties:

$$\varepsilon(\omega) = \varepsilon_\infty - \frac{\omega_\rho^2}{\omega^2 + i\omega\gamma^2} \tag{1}$$

Where ε_∞ is the relative permittivity at high frequencies, ω_ρ is the plasma frequency and γ = 1/r is the damping in time.

With this method Drude introduces equations for conductivity, resistivity and heat capacity of the metals that successfully correspond with experimental measurements. Nevertheless, it fails to describe the thermal properties of the metals [7].

3. Johnson and Christy

During the first half of the XX century various experiments were made to measure the dielectric constants of the noble metals. These experiments showed disagreement with the Drude theory in the visible and near infra-red spectrum. With the development of quantum mechanics, it was discovered that the reason of this differences was originated because Drude model does not consider the transition between bands that generate absorption of energy.

Due to the lack of a set of reliable measurements of the dielectric constants of the metals, Johnson and Christy developed a rigorous method of measuring the optical constants of noble metals with an estimated experimental error. To do so, a set of thin films of noble metals were carefully fabricated using quick evaporation and incident light in the interval 0.64 to 6.60 eV was used to measure transmittance and reflectance [5].

Johnson and Christy developed a 3-parameter method that consists of inverting the transmittance and reflection functions and the transmission of p-polarized light that enables to find the thickness of the film and the n and k parameters of the complex refraction index:

$$\mathbf{n} = n + ik \tag{2}$$

As stated, the most important optical property is the dielectric constant because it gives a description of the electronic structure of the metal and can be easily compared to the Drude theory. The dielectric constant is related to the refraction index as it follows:

$$\varepsilon(\omega) = \varepsilon_1 + i\varepsilon_2 = n^2 \tag{3}$$

Even though the optical constants measured by Johnson and Christy were published in 1972, these parameters are still referenced in current applications because of the near to the theory values in the free electron and interband range, and even when the original paper only provides optical constants in a limited range, it is possible to extrapolate using Drude model or the interband theory to get original estimated values.

SPPs are commonly described as trapped waves between metal surfaces caused by the interaction with the conductor free electrons. In this process, there is an oscillation in the electrons that is in resonance to the light wave. This resonance course between the interface charged particles and the electromagnetic field of light is what is called an SPP that propagates through the area located between a dielectric and a conductor as vanishing fields on both sides[1].

4. Demultiplexer of 12 channels

A plasmonic filter with a rectangular cavity coupled to an MIM waveguide is first proposed [9], [10] and based on this structure a demultiplexing structure for n-channels is demonstrated [4]. Based on the more recent 8-channel demultiplexing structure [10], the waveguide height, the coupling distances between cavities and the exit ports were modified in order to enhance transmittances ando more ports are considered in this structure to filter more channels. It is proposed a 12-channel demultiplexer that operates in the optical range with wavelengths between 500 and 2200 nm. The demultiplexer consists of an array of 8 cavities coupled to the main MIM waveguide. When the generation of SPPs occur in the MIM interface, the surface plasmons are created in the waveguide walls, hence the location of the exit ports to both sides of the structure. This doubles the demultiplexer efficiency.

Although it seems logical to place more than 8 cavities on the structure to filter more wavelengths, the simulation of the filter as a demultiplexer denotes that there is a limit on the exit ports that can be located as a sequential array, considering above 50% transmittances. This constraint is not caused due to the main waveguide signal decaying, as this is very low due to the materials and the size of the waveguide in nanometers. It is important to remark that every travelling wavelength that passes throw the waveguide must interact with all the exit ports to some degree. Considering this, the interacting frequencies also produce transmittances, but these are of about 20% or less, demonstrating that there is a small absorption factor for all involved wavelengths. This small absorption continues to add up throw the wave path, if it passes throw consecutive cavities, thus the energy losses are directly proportional to the exit ports quantity; this limits the total of consecutive ports that can be located in the filter to function properly. With the length of the cavities used in this model, the maximum number of cavities to work correctly are 8.

To extract the filtered wavelengths of cavities 1,2,3 and 4, two exit cavities are located in a coupling length in a position of $\frac{3}{4}$ and $\frac{1}{2}$ of the main cavity length to obtain the first and second resonant mode accordingly [6], [4]. Cavities 5,6,7, and 8 only have one exit port as shown in Figure 1. This occurs since it is physically capable to obtain the first mode of the last 4 ports as the second FP modes are located in the 800-1100 nm range, wavelengths that have been absorbed and decayed throw the first four cavities.

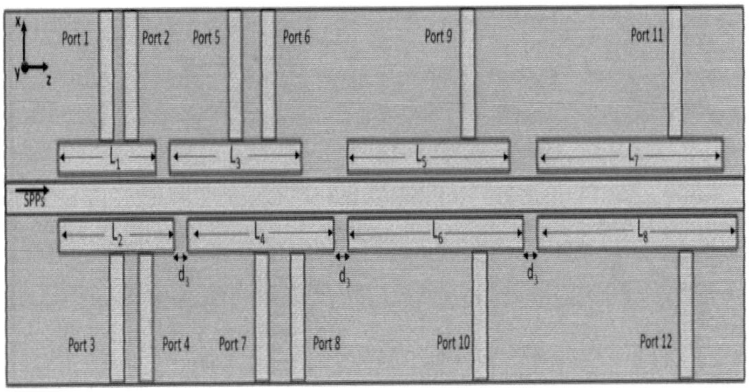

Figure 1. Theoretical MIM-based 12 channel demultiplexer.

The cavity lengths of the ports from 1 to 8 are of 281 nm, 333 nm, 381 nm, 421 nm, 451 nm, 481 nm, 531 nm, and 581 nm, respectively. The main waveguide width and the cavities are $W = 41$ nm and $W_c = 51$ nm. The coupling distance between the main waveguide and the mid ports is 11 nm and the coupling distance of each mid port to its exit port is 4 nm.

Incident light in the input port creates SPPs because of the MIM structure and propagates inside the main waveguide in the z-direction, then Farby-Perott modes are created and resonate inside the cavities. Each cavity has n resonant modes of a λm wavelength related to the cavity length L by

$$kRe(\eta_{eff})L + \Delta\theta = m\pi \tag{4}$$

where k=$\frac{2\pi}{\lambda_m}$represents the free space wave vector, $Re(\eta_{eff})$ is the cavity gap reflective index real part and $\Delta\theta$ is the reflection induced phase shift in the FP facet [11].

5. Discussions

Only the Fabry-Perott modes will resonate in each cavity so we must select the cavity length according to the wavelengths that must be filtered out of the original light beam. The selected modes of each cavity are coupled out through the exit port. In this case, the exit ports placed at $\frac{3}{4}$ of the length of the cavity extracts the first resonant mode and the one placed at $\frac{1}{2}$ extracts the second mode.

The 12 channel demultiplexer was simulated in COMSOL Multyphysics using the Drude model and the Johnson and Christy parameters. In both models we used the exact same geometry, boundary and port conditions. The only thing that changed in each study is the way the materials were modelled, for the Drude model a Domain is established then we select the Drude-Lorentz dispersion model and select all the silver domains. The Drude constants for silver are ε_∞ =3.7, $\omega\rho$=1.38 x 10^{16} rad/s and γ=2.37 x 10^{13} rad/s. Figure 2 shows the Drude model results of the demultiplexer transmission spectrum, as predicted the graph shows 12 higher than 51% transmittance peak wavelengths.

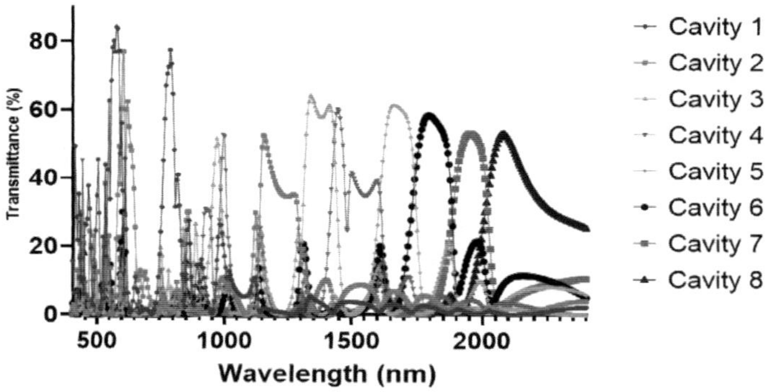

Figure 2. 12 channel MIM filter transmission spectrum. Drude model.

To use the Johnson and Christy parameters in the simulation we must select the Ag (Silver) (Johnson and Christy) form the Material Library and assign this material to the silver domain. This way COMSOL gets the optical constants of the silver to simulate the demultiplexer. The transmittance plot of all 12 exit ports of the demultiplexer using Johnson and Christy parameters are shown in Figure 3.

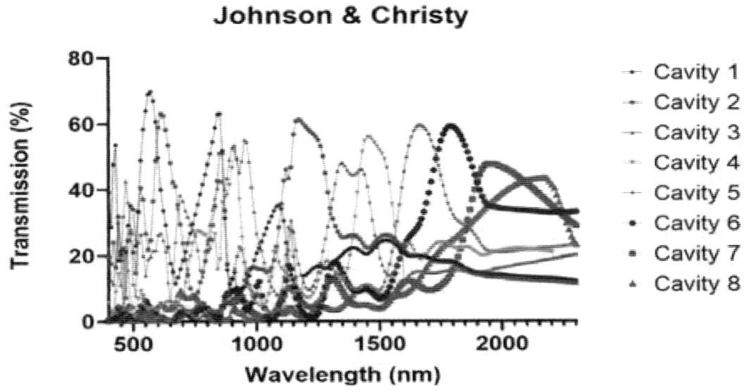

Figure 3. 12 channel MIM filter transmission spectrum. Johnson Christy model.

Note that Figure 2 and Figure 3 show 12 transmission peaks or filtered channels for 8 cavities, this is because cavities 1 to 4 have two exit ports to extract its first two modes, so the transmission curves of cavities 1-4 have 2 peaks each. But cavities 5 to 8 have only one exit port so this curves only have 1 peak each. Thereby, the transmission spectrum has 12 peaks in total, 8 from cavities 1 to 4 and 4 from cavities 5 to 8.

From both simulations above we can compare the peak or filtered wavelengths of each export. In Table 1 below we register the peak frequencies of the 12 exports and its transmissions.

Table 1. Comparison of filtered wavelengths.

Cavity	EXPORT	DRUDE Wavelength (nm)	Transmission (%)	JOHNSON AND CHRISTY WAVELENGTH(NM)	TRANSMISSION (%)
1	1	580	84.197	570	69.823
	2	790	77.483	850	63.090
2	3	605	77.004	610	63.045
	4	1155	52.529	1170	61.431
3	5	970	51.616	950	55.227
	6	1340	64.111	1350	47.974
4	7	995	52.593	910	51.912
	8	1445	60.088	1450	56.079
5	9	1670	61.249	1660	59.450
6	10	1790	58.457	1790	59.266
7	11	1950	53.083	1950	47.640
8	12	2080	53.127	2160	43.238

From Figure 2 and 3 the transmission spectrum is similar when simulating with Drude model and with Johnson and Christy parameters. With these pair of images, we can only assure that the transmission curves have the same general form but for a better understanding of the differences we should analyze some cavities individually.

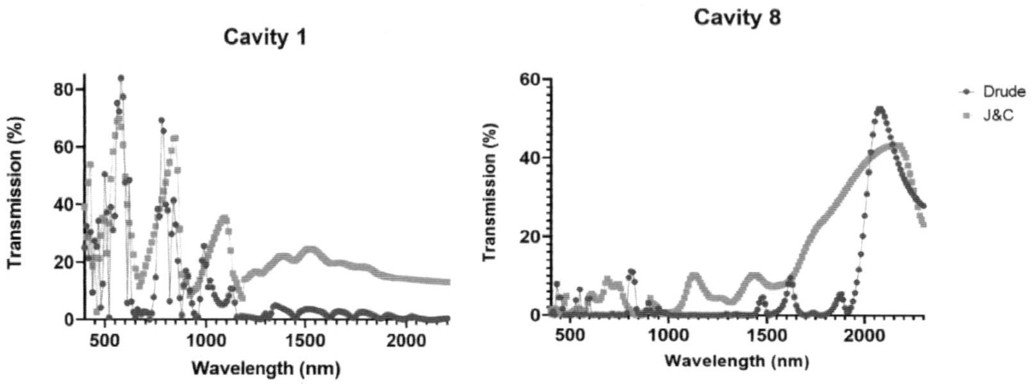

Figure 4. Transmission spectrum of cavities 1 and 8 using Drude model and Johnson and Christy parameters.

From Figure 4 we notice that both models result in transmission curves of similar shapes, and we can notice discrepancies in the interval from 400 to 700 nm due to the Drude model limitation to describe the Interband range.

6. Conclusions

Simulating SPPs nanometric devices is not an exact calculation because it does not exist a precise model for all the quantum phenomena that occurs in these interactions, but good approximations can be achieved using Drude model or Johnson and Christy.

For the demultiplexer operating in optical frequencies Drude model and Johnson and Christy offer similar transmission curves, but the wavelength peaks differ from 5 nm to 60 nm. In terms of transmission, the results are lower for the Johnson and Christy difference and the largest difference is 15% for the first mode of Cavity 1.

The most used model for other nanophotonic devices is the Drude model, this fact is due to its simplicity to use because on physics simulation software it is only necessary to specify the dielectric constant equation for the metal and input the three Drude parameters. On the contrary if the software does not have in the library the materials with its Johnson and Christy parameters it would be necessary to input all the optical constants for the selected material in the frequency range of interest.

As for computing time the Drude model approach is faster to simulate as it takes around 11 minutes to complete a frequency sweep from 400 to 2200 nm for a single cavity and for the same sweep the Johnson and Christy model takes around 21 minutes.

The simplicity and computing time mentioned above must be considered to decide which model to use but also it should be noted that the Drude model and the Johnson and Christy model provide slightly different results for plasmonic structures. In this case the small differences in transmittances between the two models can be significative. For the demultiplexer, Johnson and Christy model provide lower transmittances which would indicate that cavities 7 and 8 are not useful because they provide transmittances below 50%, while using Drude model it is proved the correct functioning of all the 8 cavities with transmittances higher than 50%.

Acknowledgments This work has been supported by project 20221182 provided by Secretaría de Investigación y Posgrado, Instituto Politécnico Nacional.

References

[1] William L Barnes. "Surface plasmon–polariton length scales: a route to sub-wavelength optics". In: *Journal of Optics A: Pure and Applied Optics* 8 (Mar. 2006), S87–S93. DOI: 10.1088/1464-4258/8/4/s06. (Visited on 01/15/2023).

[2] Dmitri K. Gramotnev and Sergey I. Bozhevolnyi. "Plasmonics beyond the diffraction limit". In: *Nature Photonics* 4 (Jan. 2010), pp. 83–91. DOI: 10.1038/nphoton.2009.282. (Visited on 01/28/2022).

[3] Eugene Hecht. *OPTics FiFTh EdiTiON GlObAl EdiTiON*. URL: https://edisciplinas.usp.br/pluginfile.php/5054148/mod_resource/content/1/Hecht-optics-5ed.pdf.

[4] Feifei Hu, Huaxiang Yi, and Zhiping Zhou. "Wavelength demultiplexing structure based on arrayed plasmonic slot cavities". In: *Opt. Lett.* 36.8 (Apr. 2011), pp. 1500–1502. DOI: 10.1364/OL.36.001500. URL: https://opg.optica.org/ol/abstract.cfm?URI=ol-36-8-1500.

[5] P. B. Johnson and R. W. Christy. "Optical Constants of the Noble Metals". In: *Physical Review B* 6 (Dec. 1972), pp. 4370–4379. DOI: 10.1103/physrevb.6.4370.

[6] Yung-Chiang Lan, Che-Jung Chang, and Peng-Hsiao Lee. "Resonant tunneling effects on cavity-embedded metal film caused by surface-plasmon excitation". In: *Opt. Lett.* 34.1 (Jan. 2009), pp. 25–27. DOI: 10.1364/OL.34.000025. URL: https://opg.optica.org/ol/abstract.cfm?URI=ol-34-1-25.

[7] Y. Li. *Chapter 1 Optical Properties of Plasmonic Materials. In Plasmonic Optics: Theory and applications.* 2017.

[8] Liu Liu, Zhanghua Han, and Sailing He. "Novel surface plasmon waveguide for high integration". In: *Opt. Express* 13.17 (Aug. 2005), pp. 6645–6650. DOI: 10.1364/OPEX.13.006645. URL: https://opg.optica.org/oe/abstract.cfm?URI=oe-13-17-6645.

[9] *Optica Publishing Group.* opg.optica.org. URL: https://opg.optica.org/oe/fulltext.cfm?uri=oe-17-9-7549&id=179265 (visited on 01/15/2023).

[10] Binfeng Yun, Guohua Hu, and Yiping Cui. "Theoretical analysis of a nanoscale plasmonic filter based on a rectangular metal–insulator–metal waveguide". In: *Journal of Physics D: Applied Physics* 43.38 (Sept. 2010), p. 385102. DOI: 10.1088/0022-3727/43/38/385102. URL: https://dx.doi.org/10.1088/0022-3727/43/38/385102.

[11] Mingzhuo Zhao et al. "Novel oscillator model with damping factor for plasmon induced transparency in waveguide systems". In: *Scientific Reports* 7 (Sept. 2017), p. 10635. DOI: 10.1038/s41598-017-11335-5. URL: https://www.nature.com/articles/s41598-017-11335-5#citeas (visited on 01/15/2023).

Identification of lithological characteristics on glacial sediments using a micromechanics approach

J Meléndez-Martínez

Instituto Mexicano del Petróleo
Eje Central Lázaro Cárdenas Norte 152, San Bartolo Atepehuacan,
Del. Gustavo A. Madero, Mexico City, Mexico. C.P. 07730.

E-mail: jaimemldz@gmail.com

Abstract. To identify the lithology present along the 1138 m length of the AND-2A drillcore; compressional wave velocity V_p and wet bulk density ρ_{WBD} data sets acquired on the drillcore are portrayed on density-velocity Rock Physics Templates (RPTs) constructed using a Self-Consistent (SC) micromechanics approach. The three most abundant minerals present in the drillcore— quartz, calcite, and mixed clays —as well as the pore fluid (brine), served as the initial values for the SC model. Results are then compared with the lithology that is identified using Gardner density-velocity cross plots. Thus, the principal lithologies found in the AND-2A drillcore have already been mentioned in the scientific works, and results from both the Gardner and SC approaches show a good correlation with these lithological features. When compared to the lithology of six specific rock specimens that were acquired at various depths along the AND-2A drillcore, our findings likewise match up well. These findings imply that the suggested SC technique may be useful for identifying lithology in boreholes where downhole physical characteristics could exist during intervals when the drillcore was not recovered in whole. Additionally, the SC approach is helpful even when physical properties are derived from measurements because lithology can occasionally be problematic to estimate from visual examination alone, and supplementary evidence from the evaluation of the physical properties could offer further understanding.

1. Introduction

Since the Cenozoic, Antarctica has experienced many ice sheets, whilst the Northern Hemisphere has experienced ongoing glaciations since the commencement of the Pliocene [1]. Consequently, sediments that were deposited by glaciers now cover large areas. People that have established in these locations, mostly in the northern hemisphere, use the sediments for water and waste deposits [11]. Therefore, understanding the various mineral lithologies related to these geological specimens is crucial for both the progress of engineering activities and for geological- evaluations. Regardless of their importance, little is known about the lithological geophysical characteristics of these sediments because they are normally found in isolated locations, therefore making it challenging to obtain the right data required to define their lithology. In this sense, in this contribution, velocity and density data from the AND-2A is used to help determine the lithology utilizing Rock Physics Templates. These RPTs are created utilizing rock physics modelling to identify fluid content and lithology and relate the elastic characteristics of geological materials to their specific lithological features [3]. Here, we make an effort to connect the elastic characteristics of the AND-2A drillcore to lithology, using ternary RPTs.

The Southern McMurdo Sound Project (SMS), an Antarctic drilling project, collected the AND-2A drillcore [5]. The drilling began over an 8.5-meter-thick sea ice podium while submerged in a 380-meter seawater pillar, and it was completed 1138 meters below the sea bottom[8][mbsf]
.

The ρ_{WBD} and V_p core log values are employed in this contribution as the physical parameters related to the AND-2A drillcore. On the other hand, the ternary diagrams are produced using SC modelling through the Sabina and Willis [2] approach. In this SC approach, it is assumed that the compound has an isotropic solid matrix inserted with n isotropic insertions or inclusions that represent the minerals and the fluid saturating the pore volume of the rock. This heterogeneous isotropic composite thus serves as a representation of the AND-2A drillcore. The physical properties of quartz, calcite, and mixed clays were used to build the RPTs that are used in this study to analyze the AND-2A drillcore. We selected such minerals since the AND-2A drillcore primarily consists of siliciclastic sediments that are cemented by calcareous minerals, such as sandstones, mudstones, and diamictite [4]. Additionally, the primary clay minerals throughout the drillcore are smectite and illite, with kaolinite and chlorite appearing less frequently. Since we also take into account the brine that saturates the rock pores; brine is therefore seen as inclusion in its own right.

2. Methodologies

2.1. Self-Consistent approach

The effective physical parameters of the AND-2A drillcore, including the shear moduli μ_{eff}, the bulk moduli κeff, and the wet bulk density ρ_{WBDeff}, are modelled using the next expressions as proposed by Sabina and Willis [2]:

$$\kappa_{eff} = \kappa_{n+1} + \sum_{r=1}^{n} \frac{\alpha_r(\kappa_r - \kappa_{n+1})}{1 + 3(\kappa_r - \kappa_{eff})/(3\kappa_{eff} + 4\mu_{eff})}$$

(1)

$$\mu_{eff} = \mu n + 1 + \sum_{r=1}^{n} \frac{\alpha_r(\mu_r - \mu_{n+1})}{1 + 6(\mu_r - \mu_{eff})(\kappa_e + 2\mu_{eff})/[5\mu_{eff}(3\kappa_{eff} + 4\mu_{eff})]} \quad (2)$$

$$\rho_{WBDeff} = \rho_{n+1} \sum_{r=1}^{n} \alpha_r(\rho_r - \rho_{n+1}) \quad (3)$$

The volume fraction, the shear modulus, the bulk modulus, and the density, for a particular inclusion, r can be characterized by α_r, $\mu_r \kappa_r$, and ρ_r where r= 1,2...,.η On the other hand, ρ_{n+1}, μ_{n+1}, and κ_{n+1} are the density, the shear modulus, and the bulk modulus for the solid matrix, respectively. If represents α_{n+1} the volume percentage of the solid matrix, then we have that

$$\sum_{r=1}^{n+1} \alpha_r = 1 \quad (4)$$

The following is a later rewrite of equation (4): $\sum_{r=1}^{n+1} \alpha_r = \alpha_r^{sol} + \alpha_r^{fl} = \alpha_r^{sol} + \phi = 1$, where α_r^{sol} and α_r^{fl} are used to denote the volume portion of the solids and of that of the fluid filling the pore space respectively. Therefore, ϕ is the percentage of the rock's voids that are saturated with liquid.

As was already indicated, in this study, calcite, quartz, and mixed clays serve as placeholders for the solid minerals, while brine serves as the saturating fluid. Whichever of these three minerals can serve as the solid matrix in the SC method. Thus, the residual two minerals, along with the brine, serve as insertions when whichever of these minerals acts as the solid matrix. Table 1 displays the physical characteristics of the minerals and the brine used as initial values in equations (1) to (3) [7]. Consequently $V_{peff}$is easily estimable as $V_{peff} = (\frac{K_{eff}+\mu_{eff}}{\rho_{WBDeff}})^{1/2}$

Table 1. The Physical characteristics of the minerals and the fluid utilized as initial data in the SC modelling [7].

Minerals and fluid	$\rho_r [gr/cm^3]$	Vp_r [km/s]	Vs_r [km/s]
Clays	2.6	3.41	1.63
Quartz	2.65	6.05	4.09
Calcite	2.71	6.64	3.44
Brine	1.02	1.52**	——

2.2. AND-2A core: physical features

The compressional wave velocities V_P and the wet bulk densities ρ_{WBD} are measured using the Geotek Multi-Sensor Core-Logger (MSCL) device as follows: conducting the AND-2A drillcore over a pair pulser-receiver P-wave Acoustic Roller Contact (ARC) transducers (with a central frequency of 230 kHz) allowed us to determine compressional transit times on the core diameter. The displacement of the acoustic transducers while passing over the drillcore provided the diameter of the core. Then dividing the measured transit times by the calculated core diameter enables the estimation of V_P profiles.

A wet bulk density profile ρ_{WBD} is then provided by a pair of 137Cs gamma-ray sources and detectors positioned crosswise the drillcore. In a nutshell, as gamma rays pass over the rock, they lose energy due to the medium's electrons' ability to disperse them. The radiation that is detected is reduced as a result of the latter. The following relationship exists between this energy loss and the electron density ρ_e corresponding to the observed gamma-ray ray:

$$\rho_e = 2\rho_{WBD}\frac{Z}{A}, \tag{5}$$

where for most rocks, the atomic number Z and the medium's molecular weight A are roughly translated into $\frac{Z}{A} = \frac{1}{2}$ [6].

3. Results and discussion

The velocity-density mineral RPTs created with the SC approach are shown in Figure 1(a). The upper triangle in these RPTs is a representation of solid rock and its vertices are named QR100% (quartz), CA100% (calcite), and CL100% (clay) whereas the vertices of the triangles modelled using brine (BR) that fills void volumes of 20% and 40% are named as BR20%+QR80%, BR20%+CA80%, and BR20%+CL80%, and BR40%+QR60% BR40%+CA60%, and BR40%+CL60%.

Figure 1(a) shows the ternary RPTs. Later, the entire AND-2A drillcore data is shown in Figure 1(b). We interpret the mineral lithotypes in terms of the locations where the minerals exhibit elastic behaviour. In this regard, we designate the siliceous lithotype, the clay-rich siliceous lithotype, and the argillaceous lithotype [11].

Because calcite is a scarce part of the AND-2A drillcore, it makes little contribution to the drillcore's reported elastic characteristics as also noted by Staudigel et al. [10]. Note that most rocks contain more than three dominant minerals, so even though RPTs try to model the elastic response of the AND-2A drillcore using calcite, quartz, and mixed clays as main minerals, data points outside the RPTs may specify intervals with the occurrence of solid minerals whose physical characteristics could be meaningfully dissimilar from calcite, quartz, and mixed clays.

However, in glacial materials, the presence of big lonestones can also lead to high ρ_{WBD} ($2.6gr/cm^3$) and V_P (≥ 4.5 km/s) [9].

The lithological analysis of the AND-2A drillcore using a cross plot of Gardner's equation is shown in Figure 2. Gardner's equation links density (ρ) and compressional velocity (V_P) as follows[13] :

$$\rho = \alpha V_P^0.25 \quad (6)$$

Coefficients typically have values of $\alpha = 1.66$ for sandstones and $\alpha = 1.75$ for shales[7]. As a result, the trend points of the siliceous lithotypes are represented by the sandstone line $\rho = 1.66V_P^0.25$, whereas the trend points for the argillaceous lithotypes are defined by the shale line $\rho = 1.75V_P^0.25$. Then, the clay-to-quartz content transition is thus represented by the clay-rich siliceous lithotype.

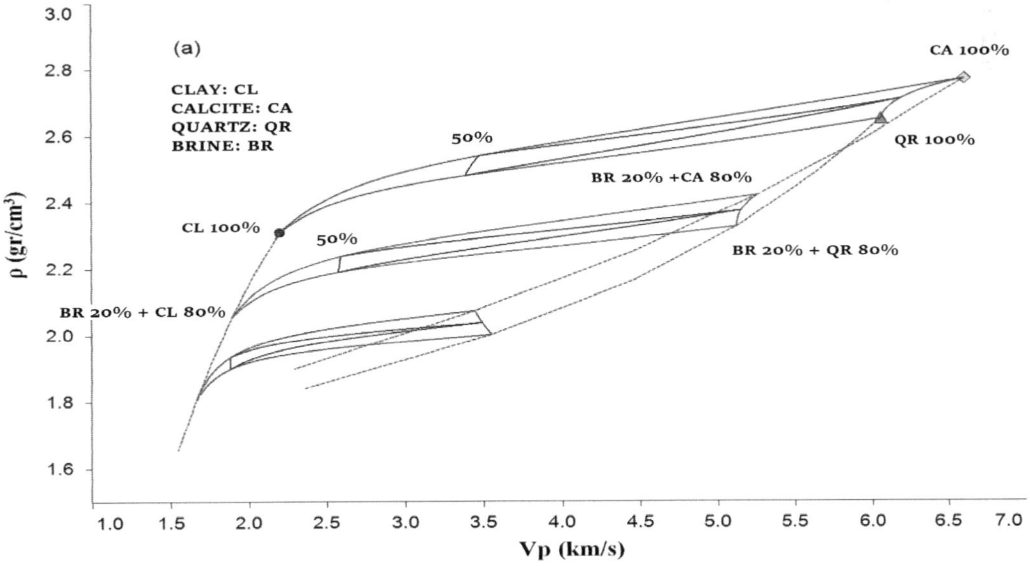

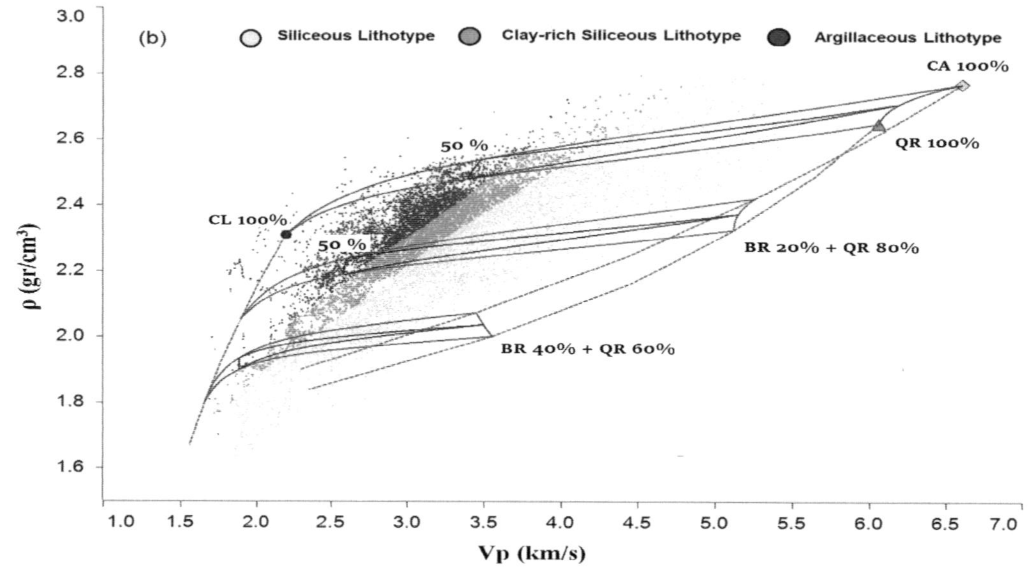

Figure 1. (a) RPTs for the density-velocity ternary minerals used to represent the elastic response of the AND-2A drillcore. (b) AND-2A drillcore´s mineral lithotype interpretation. Lithotype zones are related to the mineral elastic response.

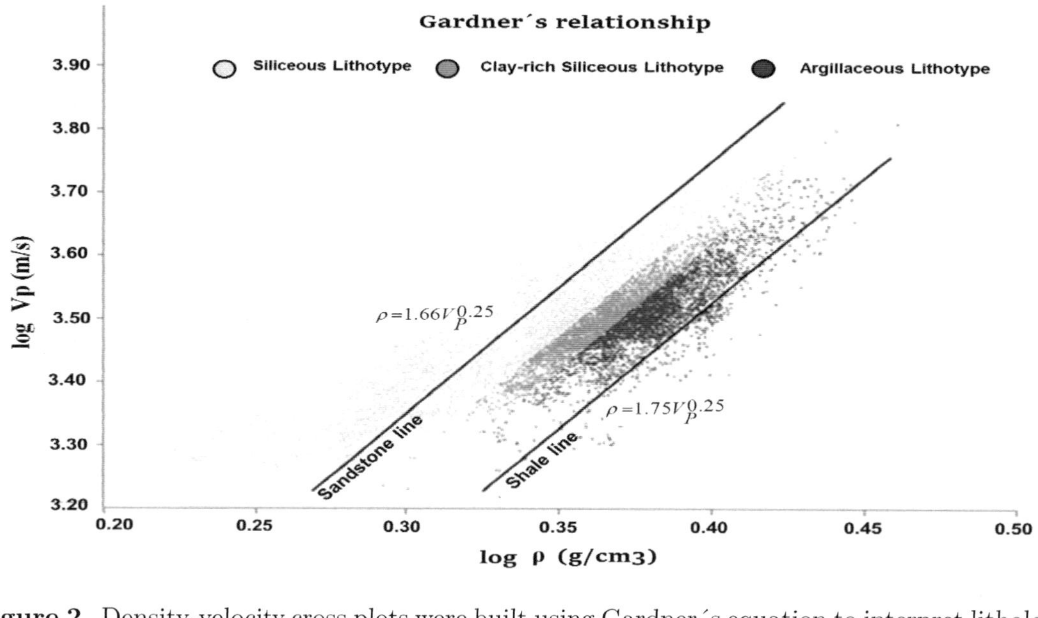

Figure 2. Density-velocity cross plots were built using Gardner´s equation to interpret lithology on the AND-2A drillcore. The sandstone line in this diagram symbolizes points with a high quartz concentration, whereas the shale line designates points with high clay content. Three mineral lithotypes are also recognized in this image: the argillaceous, clay-rich, and siliceous lithotypes.

Analysis of Figures 1(b) and 2 yielded the mineral lithotype sequences shown in Figure 3. Whereas the clay-rich siliceous lithotype is associated with clay-rich siltstones, the argillaceous lithotype in the AND-2A drillcore suggests the existence of a range of argillaceous rocks including claystone, muddy diamictite, and mudstone. On the other hand, the siliceous lithotype is connected to siliceous rocks, primarily conglomerate, sandstone, and diamictite. Results in Figure 3 are displayed in Table 2 along with the primary lithological features that match the LSUs described by Fielding et al [8]. Note that, these LSUs span over fourteen different depth intervals [mbsf] on the AND-2A. Also, LSU 8 is in turn split into 4 intervals. Nevertheless, due to a lack of information in the first LSU [0 - 37.7 mbsf], the lithological description of this unit is not considered in this contribution.

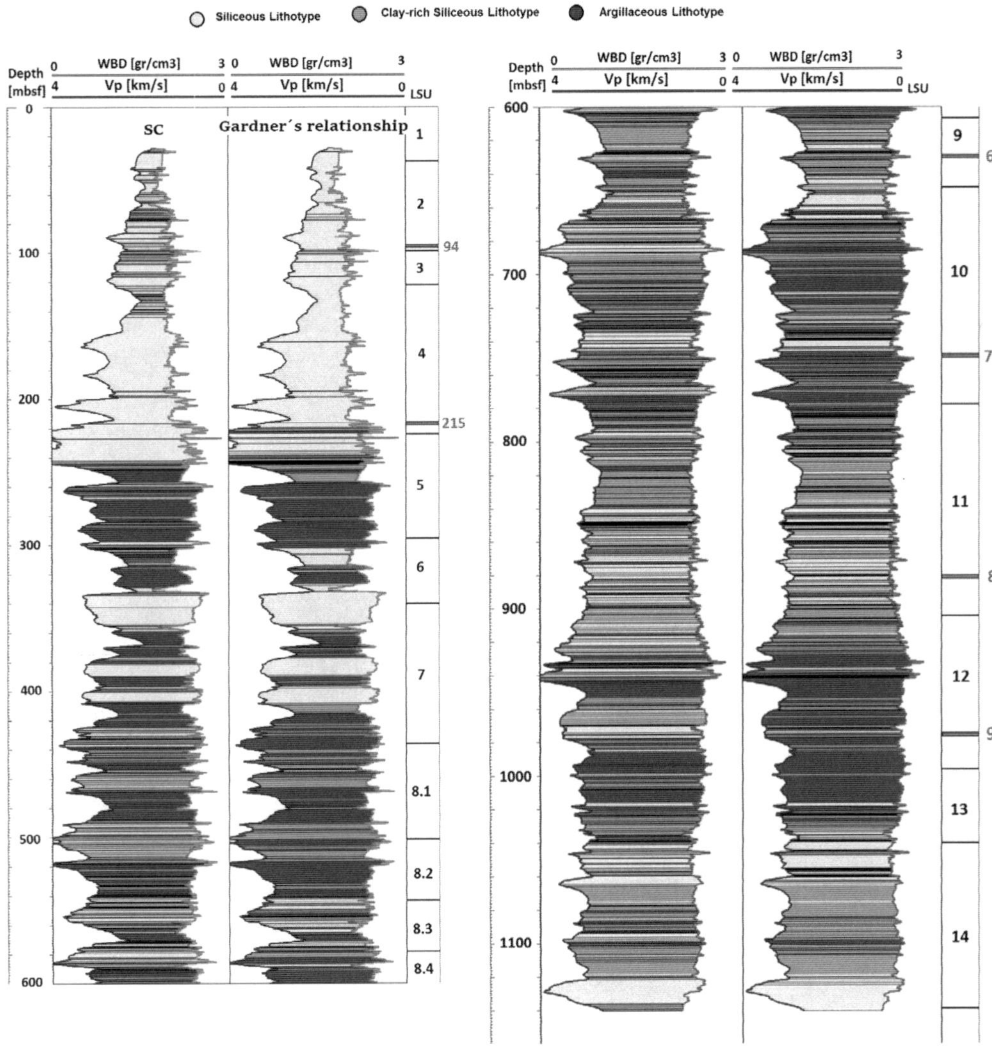

Figure 3. This figure displays the intervals associated with the 14 LSUs encompassing the AND- 2A, as well as the lithological classifications of the drillcore indicated from the analysis of Figures 1(b) and 2. Also, the depth of the six chosen samples is indicated by red markers

Table 2. The primary lithologies of the LSUs described by Fielding et al. [8] were contrasted in this Table with the lithological interpretation of the AND-2A acquired through SC and Gardner techniques.

		Lithologies		
LSU	Depth interval (mbsf)	Self-Consistent	Gardner	Fielding et al [14]
2	37.07 – 98.47	Primarily siliceous rocks with interspersed argillaceous and siltstones.	Primarily siliceous rocks with some siltstones and argillaceous rocks that are rich in clay.	Sandstone, siltstone, and sandy conglomerate interspersed with sandy diamictite
3	98.47 –122.86	Primarily siliceous rocks with interspersed argillaceous and siltstones.	Primarily siliceous rocks with some siltstones and argillaceous rocks that are rich in clay.	Sandstone and sandy conglomerate
4	122.86 –224.82	Mostly siliceous rocks with stretches of argillaceous and siltstones rich in clay.	Mostly siliceous rocks	Sandstone, siltstone, and sandy to silty
5	224.82 –296.34	Minor siliceous rocks and argillaceous rocks	Argillaceous rocks with some siliceous rocks and that are rich in clay in clay	Muddy and sandy diamictite, sandstone siltstones
6	296.34 – 339.92	Siltstones and argillaceous rocks that are rich in clay	Argillaceous rocks and siliceous rocks with the sporadic presence of siltstones rich in clay	Muddy diamictite, siltstone, claystone, and fine-grained sandstone
7	339.92 –436.18	Siliceous and argillaceous rocks with the sporadic presence of siltstones rich in clay	Siltstones and argillaceous rocks that are rich in clay.	Sandstone that contains volcanic ash and VB sandy diamictite
8.1	436.18 –502.69	Sequences of argillaceous rocks, siliceous rocks, and interbedded clay-rich siltstones.	Siltstones rich in clay interbedded with argillaceous rocks	Muddy diamictite, sandstone, and VB mudstone
8.2	502.69 –544.47	Sequences of argillaceous rocks, siliceous rocks, and clay-rich siltstones	Siltstones rich in clay interbedded with argillaceous rocks	VB mudstone, sandstone, and diamictite
8.3	544.47 –579.33	Siliceous rocks interspersed with argillaceous rocks and siltstones that are rich in clay.	Clay-rich siltstones and argillaceous rocks interspersed with siliceous rocks	Mudstone, sandy diamictite, and VB sandstone

8.4	579.33 –607.35	Clay-rich siltstones interbedded with siliceous and argillaceous rocks.	Argillaceous rocks with a negligible occurrence of limestone	Sandstone, siltstone, and VB muddy diamictite
9	607.35 – 648.74	Primarily clay-rich siltstones with a negligible amount of argillaceous and siliceous rocks.	Mostly siliceous rocks and siltstones rich in clay	siltstone and VB sandstone
10	648.74 –778.34	Sequences of argillaceous rocks, siliceous rocks, and clay-rich siltstones.	Argillaceous rocks, limestones, and siliceous rocks.	Mudstone, silty sandstone, and VB sandy diamictite
11	778.34 –904.66	Mostly argillaceous rocks and siltstones that are rich in clay.	A little number of argillaceous rocks are present among the sandstone primarily siliceous and clay-rich siltstones.	Sandy siltstone and
12	904.66 –996.69	Sequences of minor argillaceous rocks interbedded with siliceous rocks and clay-rich siltstones	Interbedding of clay-rich siltstones with argillaceous rocks. Traces of siliceous rocks are also present	Mudstone, sandstone, and diamictite
13	996.69 – 1040.28	Argillaceous rocks interspersed siliceous rocks and clay-rich siltstones	Argillaceous rocks, with siltstones and sands rich in clay	Siltstone, mudstone and sandstone
14	1040.28 - 1138.54	Presence of siliceous rocks and clay-rich siltstones	Presence of siliceous rocks and siltstones rich in clay	diamictite and sandstone

Six chosen samples recovered from the AND-2A are shown in thin slices and pictures in Figures 4 and 5. These samples were selected as being typical of the fine to coarse grain sizes identified through the three geological stages of the AND-2A Drillcore: the lower Miocene (1138.54 to 800 mbsf), the middle Miocene (800 to 223 mbsf), and the upper Miocene (223 to 0.0 mbsf)[8]. They are branded according to the depth [mbsf] at which are situated. Table 3 compares the lithological features of these samples discovered using the SC and Gardner approaches to their observed lithology. The qualitative X-ray diffraction (XRD) study (see Table 4) performed on these samples reveals that the feldspars, mica, carbonates, calcium minerals, clay minerals, and carbonates are their primary constituent minerals.

Figure 4. a) Picture and b) petrographic thin slice for sample 94: this specimen corresponds to a diamictite. Volcanic clasts and quartz minerals are attached to a silty matrix. c) Picture and d) petrographic thin slice corresponding to sample 215: coarse sandstone with a small amount of finer volcanic clasts. c) Picture and d) petrographic thin slice for sample 630: Siltstone is the lithological feature that was identified in this sample.

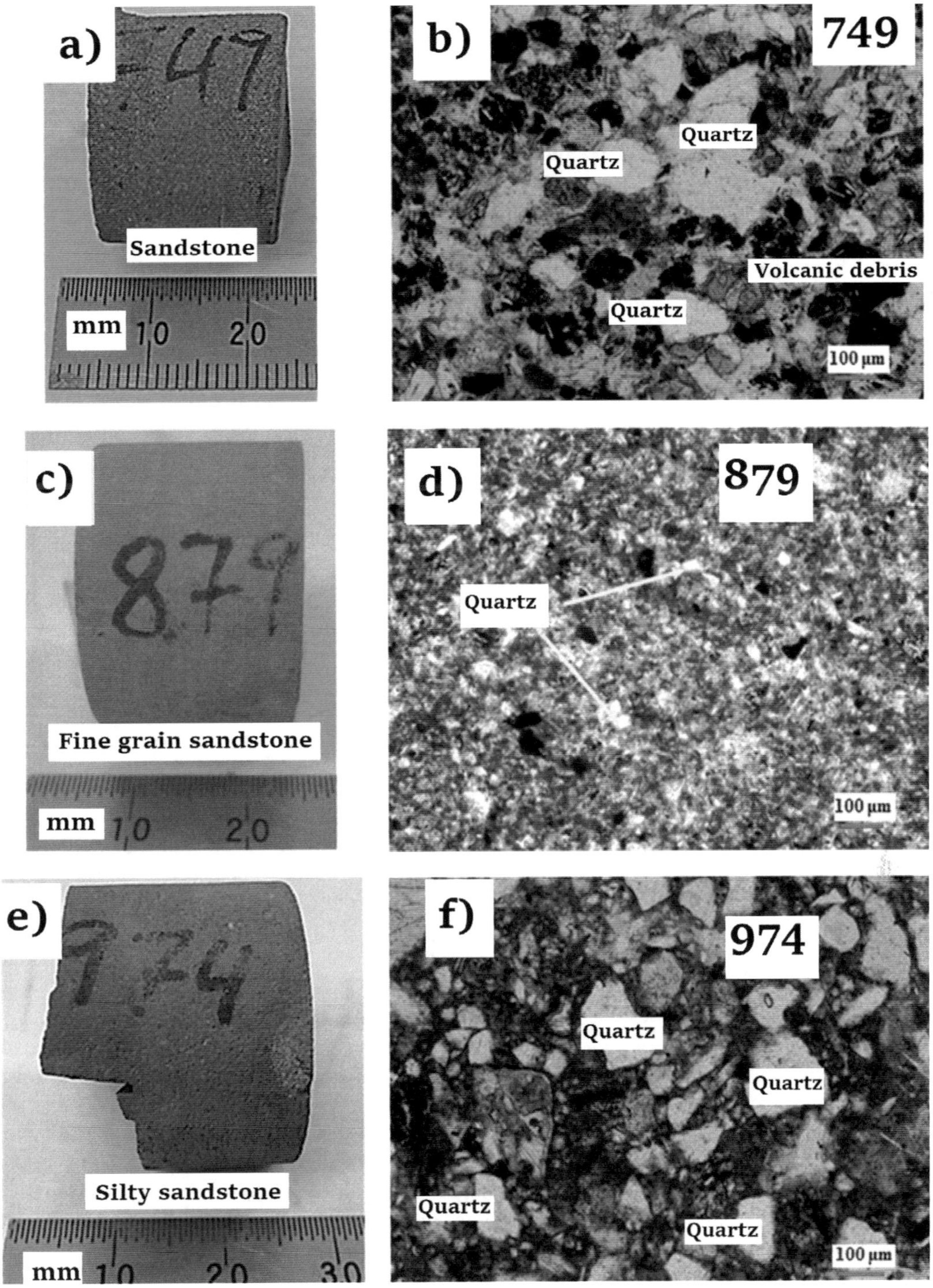

Figure 5. a) Picture and b) petrographic thin slice for sample 749: a silty sandstone has been recognized as the lithology for this sample. c) Picture and d) petrographic thin slice corresponding to sample 879: this specimen resembles a sandstone with fine grains. e) Picture and f) petrographic thin slice for sample 974: lithological features correspond to a silty sandstone.

Journal of Physics: Conference Series **2475** (2023) 012011 doi:10.1088/1742-6596/2475/1/012011

Table 3. Lithology corresponds to six chosen samples for various depths at the AND-2A obtained using visual description, the SC, and the Gardner techniques.

			Lithology	
Sample	Depth interval [mbsf]	Description	Self-Consistent	Gardner
94	94 - 96	Sandy diamictite	Siltstone rich in clay	Siliceous rock
215	215 - 215	Coarse sandstone	Siliceous rock	Siliceous rock
630	628 - 630	Sandy mudstone (siltstone)	Siltstone rich in clay	Siltstone rich in clay
749	747 - 749	Silty Sandstone	Siltstone rich in clay	Siltstone rich in clay
879	879 - 879	sandstone	Siliceous rock	Siliceous rock
974	974 - 975	Silty sandstone	Siliceous rock	Siltstone rich in clay

Table 4. Identification of minerals using XRD.

Mineral											
Sample (mbsf)	Quartz	Siderite	Anorthite	Orthoclase	Albite	Sanidine	Muscovite	Smectite	Phillipsite-Cae	Illite	Muscovite
94											
215											
630											
749											
879											
974											

We can see from Tables 2 and 3 that the six chosen samples and the described lithology in Fielding et al. [8] are in agreement with the lithology identified by the Gardner and SC approaches. As a result, LSU 2, LSU 3, and LSU 4 are all composed primarily of siliceous rocks rich in clay. Additionally, these lithologies match the lithology of samples 94 and 215. While siliceous rocks (mostly sandstones) are characteristic of LSU 7; LSU 5 and LSU 6 exhibit a change from siliceous rocks to muddy siltstones. Argillaceous rocks and siliceous rocks make up the majority of LSUs 8.1 and 8.2. The majority of the rocks in LSUs 8.3 and 8.4 are siliceous specimens. At LSU 9, a finer lithology that primarily consists of siltstone and sandstone with small mudstones that are clay-rich arises. These results are supported by the lithological description of rock specimen 630 which is siltstone containing large amounts of clay content. LSU 10 revealed an accumulation of muddy siltstones. Sample 749 is classified in Table 3 as a mixture of a siliceous rock and clay-rich siltstone. The majority of the LSU 11 is made up of muddy siltstones. The lithology of sample 879 is likewise consistent with this lithology. Interspersed sequences of siltstones and siliceous rocks with minor mudstones make up LSU 12. Particularly, the lithological description of sample 974 is consistent with the siliceous lithotype as shown in Figure 3. Muddy rocks with siltstones rich in clay are common in LSU 13. On the other hand, LSU 14 has lithologies with a coarser grain consisting of siltstones and siliceous rocks.

4. Conclusions

In this contribution, we use micromechanical SC modelling to create velocity-density RPTs for identifying mineral lithotypes on the AND-2A drillcore. The findings support Staudigel et al (2018) assertion that calcite is a rare mineral in the different lithologies present in the AND-2A. As opposed to this, the acknowledged lithologies consist of the siliceous lithotype rich in clay, the siliceous lithotype and the argillaceous lithotype. Conversely, our interpretation of Gardner's relationship is supported by the distribution of mineral lithotypes shown by the RPTs over the AND-2A drillcore. Additionally, these findings are in line with the primary lithological characteristics listed by Fielding et al. [8]. Also, exist a direct relationship between the lithological features of the six chosen samples and the predicted mineral lithotypes. A more thorough interpretation should, however, take into account petrophysical and stratigraphic investigations based on well log analysis including gamma-ray logs, resistivity logs, and neutron porosity logs, which are frequently used to distinguish lithological characteristics. However, these types of well logs are frequently not accessible. In this regard, the fundamental benefit of RPTs derived from SC modelling is the ability to distinguish mineral lithotypes only based on the elastic physical characteristics of the rock. Additionally, any other trio of minerals may be employed. This is especially helpful because scientific material on the physical characteristics of numerous minerals is easily accessible.

Acknowledgements

The author thanks Instituto Mexicano del Petróleo for funding this investigation.

References

[1] *'Late Neogene chronostratigraphy and depositional environmen.* Antarctica NZ. URL: https://adam.antarcticanz.govt.nz/nodes/view/63820 (visited on 01/15/2023).

[2] *A simple self-consistent analysis of wave propagation in particulate composites.* scholar.google.com. URL: https://scholar.google.com/citations?view_op=view_citation&hl=es&user=mNiEc-0AAAAJ&citation_for_view=mNiEc-0AAAAJ:u5HHmVD_u08C (visited on 01/15/2023).

[3] D. Close et al. "Integrated workflows for shale gas and case study results for the Horn River Basin, British Columbia, Canada". In: *The Leading Edge* 31 (May 2012), pp. 556–569. DOI: 10.1190/tle31050556.1. (Visited on 01/15/2023).

[4] C R Fielding et al. *Sequence stratigraphy of the ANDRILL AND-2A drillcore, Antarctica: A long-term, ice-proximal record of Early to Mid- Miocene climate, sea-level and glacial dynamism Palaeogeography Palaeoclimatology Palaeoecology.* 2011.

[5] F. Florindo et al. "ANDRILL's Success During the 4th International Polar Year". In: *Scientific Drilling* 6 (2008), pp. 29–31. DOI: 10.2204/iodp.sd.6.03.2008. URL: https://sd.copernicus.org/articles/6/29/2008/.

[6] George V. Keller. "A Practical Introduction to Borehole Geophysics". In: *Eos, Transactions American Geophysical Union* 69 (1988), p. 850. DOI: 10.1029/88eo01111. (Visited on 01/15/2023).

[7] Gary Mavko, Tapan Mukerji, and Jack Dvorkin. *The Rock Physics Handbook: Tools for Seismic Analysis of Porous Media.* 2nd ed. Cambridge University Press, 2009. DOI: 10.1017/CBO9780511626753.

[8] Jaime Meléndez Martínez, Rubén Nicolás López, and Oscar C Valdiviezo. "Mineral lithotype identification on the andrill AND-2A drillcore, antarctica by using ternary mineral rock physics templates built from a self-consistent approach". In: *Geofísica Internacional* 61 (Jan. 2022), pp. 40–54. DOI: 10.22201/igeof.00167169p.2022.61.1.2126. (Visited on 03/31/2022).

[9] Frank Niessen, R. D. Jarrard, and C. Bücker. "Log-based physical properties of the CRP-1 core, Ross Sea, Antarctica". In: *Terra Antartica* 5 (1998), pp. 299–310. URL: https://epic.awi.de/id/eprint/2226/ (visited on 01/15/2023).

[10] P T Staudigel et al. *Cryogenic brines as diagenetic fluids: Reconstructing the diagenetic history of the Victoria Land Basin using clumped isotopes Geochimica et Cosmochimica.* 2018.

[11] D. A. Stephenson, A. H. Fleming, and D. M. Mickelson. "Glacial deposits". In: *Hydrogeology.* Geological Society of America, Jan. 1988. ISBN: 9780813754673. DOI: 10.1130/DNAG-GNA-02.301. URL: https://doi.org/10.1130/DNAG-GNA-02.301.

AUTHOR INDEX